Advanced Welding Processes

New Manufacturing Processes and Materials Series

Series Editors: **John Wood**, Nottingham University, UK
Diran Apelian, Worcester Polytechnic Institute, USA

Already published

Powder Metallurgy: The Process and its Products
G Dowson

Hot Isostatic Processing
H V Atkinson and B A Rickinson

Sheet Metal Forming
R Pearce

New Manufacturing Processes and
Materials Series

Advanced Welding Processes

John Norrish

*Head of Welding Group, School of
Industrial and Manufacturing Science,
Cranfield Institute of Technology*

Institute of Physics Publishing
Bristol, Philadelphia and New York

British Library Cataloguing in Publication Data

A catalogue record for this book is available from the British Library.

    ISBN 0–85274–325–4 (hbk)
           0–85274–326–2 (pbk)

Library of Congress Cataloging-in-Publication Data are available

Series Editors: **John Wood**, Nottingham University, UK
                Diran Apelian, Worcester Polytechnic Institute, USA

Published by IOP Publishing Ltd, a company wholly owned by the Institute of Physics, London

IOP Publishing Ltd
Techno House, Redcliffe Way, Bristol BS1 6NX, UK
335 East 45th Street, New York, NY 10017-3483, USA
US Editorial Office: IOP Publishing Inc., The Public Ledger Building, Suite 1035, Independence Square, Philadelphia, PA 19106

Typeset by Mathematical Composition Setters Ltd, Wiltshire
Printed in the UK by J. W. Arrowsmith Ltd, Bristol

FOR LUKE

In whose memory proceeds due to the author will be donated
to the Foundation for Conductive Education, PO Box 363,
Birmingham, UK. (Registered charity no 295873.)

Contents

Series Editors' Foreword

It is becoming widely recognized that courses in Materials Science and Engineering at colleges, polytechnics and universities must be seen in the general context of manufacturing. Conversely the lip-service often paid by engineering students to materials is now seen as an Achilles' heel in the training of personnel who have to make design decisions. There is still a necessity for research and teaching in the fundamental physics and chemistry of materials (referred to as materials science and covered in physics, chemistry and biological sciences, in addition to metallurgy and materials science departments) but there is a genuine gap when considering materials design and engineering. The aim of this series is to provide state-of-the-art books which address the interface between materials manufacturing and design. They are not intended to be treatises on the fundamentals of materials, but rather designed for students who appreciate the need to quantify materials performance in terms of fundamental parameters but need to use them in, and for, real situations and applications.

Educationalists in universities are extremely concerned that graduates in manufacturing-related courses should be made aware of the implications of the material, the manufacturing process, and the design criteria on the final object. It has to be said that this area is covered extremely well in a number of industrial countries by the use of a team approach to the design process. In the UK, recent surveys have demonstrated that there is an appalling lack of knowledge about materials and processes among design engineers in many industries. It is this situation that is now blamed for much of the lack of a competitive edge in the application of basic science to real life problems. The authors in this series have been asked to tackle both materials and manufacturing processes in this context. The books are intended to be of interest to students on engineering, and specifically manufacturing engineering, courses. In the main they will be written by people who have considerable experience of the industrial side of engineering and as such will not require high levels of mathematical skills. In addition, they are intended to be useful as way-in texts for practising engineers who need to make themselves acquainted with a new field or new material in a context which is relevant to them. The success of the series will be measured by the extent to which design engineers and production engineers are influenced to make new products in an innovative way.

The Editors

Preface

Welding has traditionally been regarded as a craft rather than a technological manufacturing process. This reputation has not been helped by the dependence of conventional joining techniques on highly skilled manual operators and the relatively high cost and poor reproduceability associated with many welding processes.

Developments in welding technology have in the past been based largely on metallurgical research, which has enabled a wide range of materials to be joined, adequate joint properties to be maintained and the integrity of welded joints to be controlled. Whilst further work is still required in the materials research area, in particular to enable advanced materials to be joined effectively, much of the recent effort has been devoted to improved understanding of the basic processes, consumable and equipment development and control and automation.

This book outlines the common production welding processes, discusses developments in consumables and processes and covers in some detail the developments in electronic power regulation, computer control, automation and process monitoring. It attempts to draw together the latest work in each area and indicates how the advances in welding technology may be used to produce cost-effective joints of appropriate quality.

John Norrish
Cranfield Institute of Technology

Acknowledgments

I would like to acknowledge the help, advice, encouragement and goading which have been generously provided by many friends and colleagues in the welding industry. Also my students, who have taught me much and whose work is reported widely in the following pages. Thanks are also due to the editor of *Welding and Metal Fabrication* for permission to reproduce some of the material which I have previously published in that journal, and to the many companies who have supplied illustrations.

Last, but by no means least, I would like to thank my long-suffering family for their forbearance during the protracted gestation of this work.

1 An Introduction to Welding Processes

1.1 Introduction

Welding and joining are essential for the manufacture of a range of engineering components, which may vary from very large structures such as ships and bridges, to very complex structures such as aircraft engines, or miniature components for microelectronic applications.

1.1.1 Joining processes

The basic joining processes may be subdivided into:

(i) mechanical joining;
(ii) adhesive bonding;
(iii) brazing and soldering;
(iv) welding.

A large number of joining techniques are available, and in recent years significant developments have taken place, particularly in the adhesive bonding and welding areas. Existing welding processes have been improved and new methods of joining have been introduced. The proliferation of techniques which have resulted makes process selection difficult and may limit their effective exploitation.

The aim of this book is to provide an objective assessment of the most recent developments in welding process technology in an attempt to ensure that the most appropriate welding process is selected for a given application.

This chapter will introduce some of the basic concepts which need to be considered and highlight some of the features of traditional welding methods.

1.1.2 Classification of welding processes

Several alternative definitions are used to describe a weld, for example:

A union between two pieces of metal rendered plastic or liquid by heat or pressure or both. A filler metal with a melting temperature of the same order of that of the parent metal may or may not be used [2].

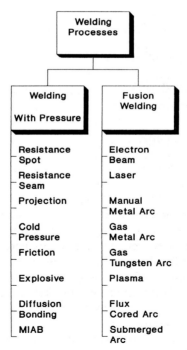

Figure 1.1 Important welding processes.

or alternatively:

A localized coalescence of metals or nonmetals produced either by heating the materials to the welding temperature, with or without the application of pressure, or by the application of pressure alone, with or without the use of a filler metal [1].

Based on these definitions welding processes may be classified into those which rely on the application of pressure and those which use elevated temperatures to achieve the bond. A chart illustrating the derivation of welding processes on this basis has been published in British Standard BS499 [2] and this is summarized in appendix 1. Many of the forty or so processes referred to in this classification are of little industrial importance but a small number of these processes are used extensively. Some of the most important processes are shown in figure 1.1.

1.2   Conventional Welding Processes

A brief description of the most common processes, their applications and limitations is given below. The more advanced processes and their developments are dealt with in more detail in the remaining chapters.

1.2.1   Welding with pressure

(a)   *Resistance welding*
The resistance welding processes are commonly classified as pressure welding processes although they involve fusion at the interface of the material being joined. Resistance spot, seam and projection welding rely on a similar mechanism. The material to be joined is clamped between two electrodes and a high current is applied (figure 1.2). Resistance heating at the contact surfaces causes local melting and fusion. High currents (typically 10 000 A) are applied for short durations and pressure is applied to the electrodes prior to the application of current and for a short time after the current has ceased to flow.

Accurate control of current amplitude, pressure and weld cycle time are required to ensure that consistent weld quality is achieved but some variation may occur due to changes in the contact resistance of the material, electrode wear, magnetic losses or shunting of the current through previously formed spots. These 'unpredictable' variations in process performance have led to the practice of increasing the number of welds from the design requirement to give some measure of protection against poor individual weld quality. To improve this situation significant developments have been made in resistance monitoring and control, these allow more efficient use of the process and the techniques available are described in Chapter 10.

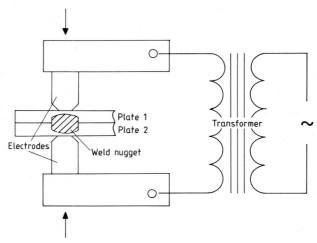

Figure 1.2   Resistance welding system.

Features of the basic resistance welding process include:

(i) the process requires relatively simple equipment;
(ii) it is easily and normally automated;
(iii) once the welding parameters are established it should be possible to produce repeatable welds for relatively long production runs.

The major applications of the process have been in the joining of sheet steel in the automotive and white-goods manufacturing industries.

(b) *Cold pressure welding*

If sufficient pressure is applied to the cleaned mating surfaces to cause substantial plastic deformation the surface layers of the material are disrupted, metallic bonds form across the interface and a cold pressure weld is formed [3].

The main characteristics of cold pressure welding are:

(i) the simplicity and low cost of the equipment;
(ii) the avoidance of thermal damage to the material;
(iii) most suitable for low-strength (soft) materials.

The pressure and deformation may be applied by rolling, indentation, butt welding, drawing or shear welding techniques. In general the more ductile materials are more easily welded.

This process has been used for electrical connections between small-diameter copper and aluminium conductors using butt and indentation techniques. Roll bonding is used to produce bimetallic sheets such as Cu/Al for cooking utensils, Al/Zn for printing plates and precious-metal contact springs for electrical applications.

(c) *Friction welding*

In friction welding a high temperature is developed at the joint by the relative motion of the contact surfaces. When the surfaces are softened a forging pressure is applied and the relative motion is stopped (figure 1.3). Material is extruded from the joint to form an upset.

The process may be divided into several operating modes in terms of the means of supplying the energy:

(1) *Continuous drive*: in which the relative motion is generated by direct coupling to the energy source. The drive maintains a constant speed during the heating phase.

(2) *Stored energy*: in which the relative motion is supplied by a flywheel which is disconnected from the drive during the heating phase.

The process may also be classified according to the type of motion as shown in figure 1.4. Rotational motion is the most commonly used, mainly for round components where angular alignment of the two parts is not critical. If it is required to achieve a fixed relationship between the mating parts angular oscillation may be used and for non-circular components the linear and orbital techniques may be employed.

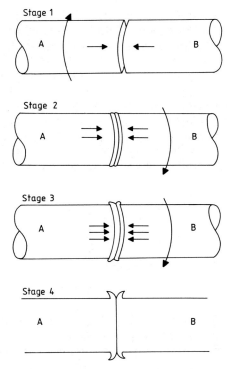

Figure 1.3 Friction welding. Stage 1: A fixed, B rotated and moved into contact with A. Stage 2: A fixed, B rotated under pressure, interface heating. Stage 3: A fixed, forge pressure applied. Stage 4: relative motion stopped, weld formed.

Features of the process include:

(i) one-shot process for butt welding sections;
(ii) suitable for dissimilar metals;

(iii) short cycle time;
(iv) most suited to circular sections;
(v) robust and costly equipment may be required.

The process is commonly applied to circular sections, particularly in steel, but it may also be applied to dissimilar metal joints such as aluminium to steel or even ceramic materials to metals. Early applications of the process included the welding of automotive stub axles but the process has also been applied to the fabrication of high-quality aero-engine parts [4], duplex stainless steel pipe for offshore applications [5] and nuclear components [6].

Recent developments of the process include the joining of metal to ceramics [7], the use of the process for stud welding in normal ambient conditions and underwater, and the use of the process for surfacing [8]. The linear technique has recently been successfully demonstrated on titanium alloy welds having a weld area of 250 mm^2 using an oscillation frequency of 25 kHz, 110 N mm^{-2} axial force and an oscillation amplitude of ± 2 mm [9].

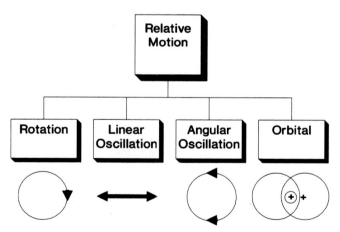

Figure 1.4   Friction welding: classification by motion type.

(d)   *Diffusion bonding*
In diffusion bonding the mating surfaces are cleaned and heated in an inert atmosphere. Pressure is applied to the joint and local

plastic deformation is followed by diffusion during which the surface voids are gradually removed [10]. The components to be joined need to be enclosed in a controlled atmosphere and the process of diffusion is time and temperature dependent. In some cases an intermediate material is placed between the abutting surfaces to form an interlayer.

Significant features of the process are:

(i) suitable for joining a wide range of materials;
(ii) one-shot process;
(iii) complex sections may be joined;
(iv) vacuum or controlled atmosphere required;
(v) prolonged cycle time.

The process can, however, be used for the joining of complex structures which require many simultaneous welds to be made.

(e) *Explosive welding*

In explosive welding the force required to deform the interface is generated by an explosive charge. In the most common application of the process two flat plates are joined to form a bimetallic structure. An explosive charge is used to force the upper or 'flier' plate on to the baseplate in such a way that a wave of plastic material at the interface is extruded forward as the plates join (figure 1.5). For large workpieces considerable force is involved and care is required to ensure the safe operation of the process. Features of the process include:

(i) one-shot process—short welding time;
(ii) suitable for joining large surface areas;
(iii) suitable for dissimilar thickness and metals joining;
(iv) careful preparation required for large workpieces.

The process may also be applied for welding heat exchanger tubes to tube plates or for plugging redundant or damaged tubes.

(f) *Magnetically impelled arc butt welding* (MIAB)

In MIAB welding a magnetic field generated by an electromagnet is

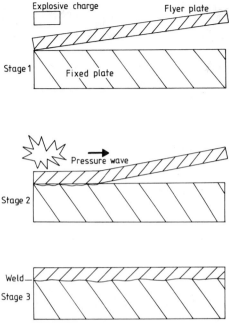

Figure 1.5   Explosive welding.

used to move an arc across the joint surfaces prior to the application of pressure [11] (figure 1.6). Although the process produces a similar weld to friction welding it is possible to achieve shorter cycle times and relative motion of the parts to be joined is avoided. Features of the process are:

  (i) one-shot process;
  (ii) suitable for butt welding complex sections;
  (iii) shorter cycle time than friction welding.

The process has been applied fairly widely in the automotive industry for the fabrication of axle cases and shock absorber housings in tube diameters from 10 to 300 mm and thicknesses from 0.7 to 13 mm [12].

Stage 1

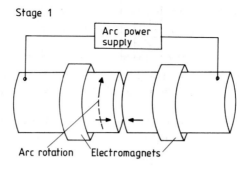

Arc rotation   Electromagnets

Stage 2

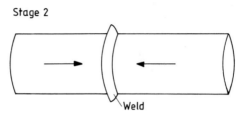

Weld

Figure 1.6 MIAB welding. Stage 1: rotating arc heats end faces. Stage 2: forge pressure applied.

1.2.2 Fusion welding

(a) Gas tungsten arc welding (GTAW)

In the gas tungsten arc welding process† the heat generated by an arc which is maintained between the workpiece and a non-consumable tungsten electrode is used to fuse the joint area. The arc is sustained in an inert gas which serves to protect the weld pool and the electrode from atmospheric contamination (figure 1.7).

The process has the following features:

  (i) it is conducted in a chemically inert atmosphere;
  (ii) the arc energy density is relatively high;
  (iii) the process is very controllable;

† The GTAW process is also known as tungsten inert gas (TIG) in most of Europe, WIG (wolfram inert gas) in Germany, and is still referred to by the original trade names Argonarc or Heliarc welding in some countries.

(iv) joint quality is usually high;

(v) deposition rates and joint completion rates are low.

The process may be applied to the joining of a wide range of engineering materials including stainless steel, aluminium alloys and reactive metals such as titanium. These features of the process lead to its widespread application in the aerospace, nuclear reprocessing and power generation industries as well as in the fabrication of chemical process plant, food processing and brewing equipment.

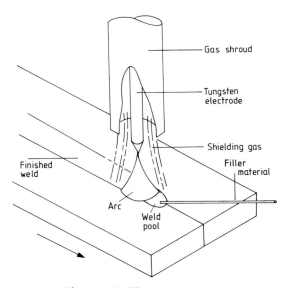

Figure 1.7   The GTAW process.

(b) *Shielded metal arc welding (SMAW)*

Shielded metal arc welding† has for many years been one of the most common techniques applied to the fabrication of steels. The process uses an arc as the heat source but shielding is provided by gases generated by the decomposition of the electrode coating

† The SMAW process is also known as MMA (manual metal arc welding) in Europe and is still referred to as *Stick* welding in the UK fabrication industry.

material and by the slag produced by the melting of mineral constituents of the coating (figure 1.8). In addition to heating and melting the parent material the arc also melts the core of the electrode and thereby provides filler material for the joint. The electrode coating may also be used as a source of alloying elements and additional filler material. The flux and electrode chemistry may be formulated to deposit wear- and corrosion-resistant layers for surface protection.

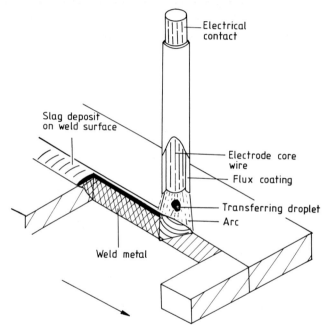

Figure 1.8 Shielded metal arc welding (or manual metal arc (MMA) welding).

Significant features of the process are:

(i) equipment requirements are simple;
(ii) a large range of consumables are available;
(iii) the process is extremely portable;
(iv) the operating efficiency is low;
(v) it is labour intensive.

For these reasons the process has been traditionally used in structural steel fabrication, shipbuilding and heavy engineering as well as for small batch production and maintenance.

(c) *Submerged arc welding (SAW)*

Submerged arc welding is a consumable electrode arc welding process in which the arc is shielded by a molten slag and the arc atmosphere is generated by decomposition of certain slag constituents (figure 1.9). The filler material is a continuously fed wire and very high melting and deposition rates are achieved by using high currents (e.g. 1000 A) with relatively small-diameter wires (e.g. 4 mm).

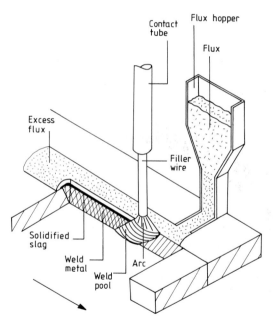

Figure 1.9 Submerged arc welding.

The significant features of the process are:

(i) high deposition rates;
(ii) automatic operation;

(iii) no visible arc radiation;
(iv) flexible range of flux/wire combinations;
(v) difficult to use positionally;
(vi) best for thicknesses above 6 mm.

The main applications of submerged arc welding are on thick-section plain carbon and low-alloy steels and it has been used on power generation plant, nuclear containment, heavy structural steelwork, offshore structures and shipbuilding. The process is also used for high-speed welding of simple geometric seams in thinner sections, for example in the fabrication of pressure containers for liquefied petroleum gas. Like shielded metal arc welding, with suitable wire/flux combinations the process may also be used for surfacing.

(d) Gas metal arc welding (GMAW)

Gas metal arc welding† uses the heat generated by an electric arc to fuse the joint area. The arc is formed between the tip of a consumable, continuously fed filler wire and the workpiece and the entire arc area is shielded by an inert gas. The principle of operation is illustrated in figure 1.10.

Some of the more important features of the process are summarized below:

(i) low heat input (compared with SMAW and SAW);
(ii) continuous operation;
(iii) high deposition rate;
(iv) no heavy slag—reduced post-weld cleaning;
(v) low hydrogen—reduces risk of cold cracking.

Depending on the operating mode of the process it may be used at low currents for thin sheet or positional welding.

The process is used for joining plain carbon steel sheet from 0.5 to 2.0 mm thick in the following applications: automobile bodies, exhaust systems, storage tanks, tubular steel furniture, heating and ventilating ducts. The process is applied to positional welding of

† Gas metal arc welding is also known as metal inert gas (MIG) or metal active gas (MAG) welding in Europe. The terms semi-automatic or CO_2 welding are sometimes used but are less acceptable.

thicker plain carbon and low alloy steels in the following areas: oil pipelines, marine structures and earth-moving equipment. At higher currents high deposition rates may be obtained and the process is used for downhand and horizontal–vertical welds in a wide range of materials—applications include earth-moving equipment, structural steelwork (e.g. I-beam prefabrication), weld surfacing with nickel or chromium alloys, aluminium alloy cryogenic vessels and military vehicles.

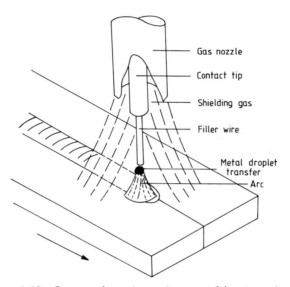

Gas nozzle

Contact tip

Shielding gas

Filler wire

Metal droplet transfer

Arc

Figure 1.10   Gas metal arc (GMAW) or metal inert gas (MIG) welding.

(e)   *Plasma welding*
Plasma welding uses the heat generated by a constricted arc to fuse the joint area, the arc is formed between the tip of a non-consumable electrode and either the workpiece or the constricting nozzle (figure 1.11). A wide range of shielding and cutting gases are used depending on the mode of operation and the application.

In the normal transferred arc mode the arc is maintained between the electrode and the workpiece; the electrode is usually the cathode and the workpiece is connected to the positive side of the power supply. In this mode a high energy density is achieved and the process may be used effectively for welding and cutting.

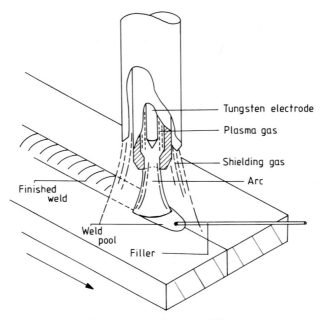

Figure 1.11   Plasma welding.

The features of the process depend on the operating mode and the current and will be described further in Chapters 5 and 7, but in summary the plasma process has the following characteristics:

(i) good low-current arc stability;
(ii) improved directionality compared with TIG;
(iii) improved melting efficiency compared with TIG;
(iv) possibility of keyhole welding.

These features of the process make it suitable for a range of applications including the joining of very thin materials, the encapsulation of electronic components and sensors, and high-speed longitudinal welds on strip and pipe.

(f) *Electron beam welding*

A beam of electrons may be accelerated by a high voltage to provide a high-energy heat source for welding (see chapter 8). The power density of electron beams is high (10^{10} to 10^{12} $W\,m^{-2}$) and keyhole welding is the normal operating mode. The problem of power dissipation when the electrons collide with atmospheric gas molecules is usually overcome by carrying out the welding operation in a vacuum. Features of the process include:

(i) very high energy density;
(ii) confined heat source;
(iii) high depth to width ratio of welds;
(iv) normally requires a vacuum;
(v) high equipment cost.

Applications of electron beam welding have traditionally included welding of aerospace engine components and instrumentation, but it may be used on a wide range of materials when high-precision and very deep penetration welds are required.

(g) *Laser welding*

The laser may be used as an alternative heat source for fusion welding (see Chapter 7). The focused power density of the laser can reach 10^{10} or 10^{12} $W\,m^{-2}$ and welding is often carried out using the 'keyhole' technique. Significant features of laser welding are:

(i) very confined heat source at low power;
(ii) deep penetration at high power;
(iii) reduced distortion and thermal damage;
(iv) out-of-vacuum technique;
(v) high equipment cost.

These features have led to the application of lasers for microjoining of electronic components, but the process is also being applied to the fabrication of automotive components and precision machine tool parts in heavy section steel.

1.3   Summary

A wide range of welding processes is available and their suitability for a given application is determined by the inherent features of the process. Specific process developments and advances in automation and process monitoring, which may be used to enhance most welding systems, will be discussed in the following chapters.

2 Advanced Process Development Trends

2.1 Introduction

The primary incentive for welding process development is the need to improve the total cost effectiveness of joining operations in manufacturing industry. However, other factors may influence the requirement for new processes. Recently, concern over the safety of the welding environment and the potential shortage of skilled technicians and operators in many countries have become important considerations.

Many of the traditional welding techniques described in Chapter 1 are regarded as costly and hazardous and it is possible to improve both of these aspects significantly by employing some of the advanced process developments described in the following chapters. The background to some of the more significant developments and current trends in the application of advanced processes are discussed below.

2.1.1 Cost effectiveness

The cost of producing a welded joint is the sum of costs associated with labour, materials, power and capital plant depreciation. The total cost of welding operations in western economies is largely governed by the cost of labour, and in many traditional welding processes this can account for 70% to 80% of the total. This is illustrated schematically in figure 2.1.

In the past it seems to have been assumed that the cost effectiveness of welding processes was totally dependent on deposition rate. Processes which gave increased deposition rate were sought and developed but many of these techniques have failed to achieve the expected cost savings and their application has been restricted. Deposition rate is also an inappropriate criterion against which to assess the single-shot techniques such as friction welding or autogenous processes such as laser welding.

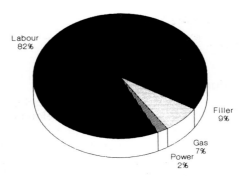

Figure 2.1   Welding costs—positional GMAW weld.

It is now clear that many additional factors must be considered when the cost effectiveness of a process is being assessed; of these the following are probably the most important:

(i) control of joint quality;
(ii) joint design;
(iii) the operating efficiency;
(iv) equipment and consumable costs.

(a) *Deposition rate*
Deposition rate may be used to compare the basic cost of the consumable electrode welding processes as shown in figure 2.2 [13]. The higher the deposition rate the shorter the potential welding cycle and this usually results in savings, predominantly in labour costs. Any appraisal of the total cost must, however, include the factors listed above.

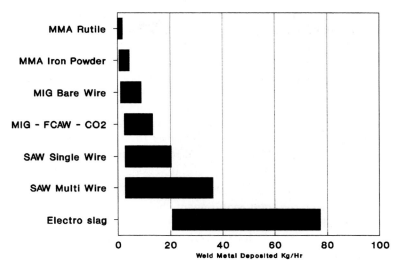

Figure 2.2 Deposition rates for a range of consumable electrode processes (after Smith [13]).

(b) *Control of joint quality*

Traditional welding processes are controlled by a large number of interrelated operating parameters and the joint quality often depends on the optimization of these parameters as well as the careful control of pre-weld and post-weld treatments. In order to ensure repeatable joint quality the operating parameters derived from a combination of established 'rules' and welding trials are defined for each joint in the form of a welding procedure [14]. For critical structural joints this welding procedure and the operator may require formal approval by a certifying authority. This process of procedure generation and qualification is both time consuming and costly and once a procedure has been established the additional cost involved in adopting a new process may be prohibitive unless the cost of requalification can be recovered from the potential savings.

The success of this control technique also depends on ensuring that the predetermined procedure is actually followed in production; this in turn means monitoring the performance of the equipment used and ensuring that the operator adheres to the original technique. Unfortunately this is not always the case and additional costs are often incurred in post-weld inspection and weld repair.

The development of techniques which enable the welding process to be controlled more effectively should have a significant impact on costs. The use of more tolerant consumables, more repeatable equipment and processes, automation, on-line monitoring and real-time control systems all contribute to improved overall process control. In addition there is renewed interest in the use of modelling and parameter prediction techniques to enable the optimum welding parameters to be established for a given welding situation.

(c) *Joint design*

Over-specifying the joint requirements has a marked effect on the cost of welding; in the case of a simple fillet weld a 1 mm increase in the specified leg length can increase the cost by 45% as shown in figure 2.3.

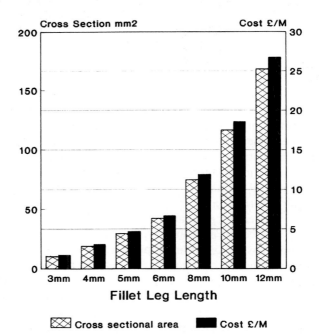

Figure 2.3 Effect of joint design (fillet leg length) on cost of weld.

The choice of a specific joint design can automatically preclude the use of the most cost-effective process; for example limited access or complex joint profiles may limit the process choice and it is important for the designer to understand the limitations of the joining process to avoid these restrictions. Conversely the selection of an appropriate process may reduce both joint preparation costs and joint completion time. In general the joint completion time is related to the required weld metal volume and it can be seen from figure 2.4 that this will vary significantly depending on the joint design. For example, using the electron beam process a butt joint in 20 mm thick steel will be completed more quickly than the equivalent MIG weld which will require a 50° to 60° included angle to enable satisfactory access.

Process developments which require low weld metal volume and limited joint preparation are therefore likely to be more cost effective.

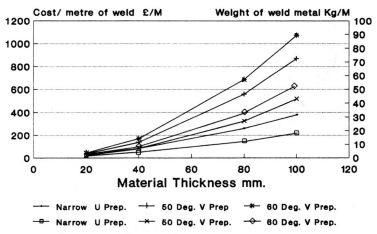

Figure 2.4 Effect of fillet geometry on weld cost.

(d) Operating efficiency

The operating efficiency of welding processes is usually expressed as the 'operating factor' † which is the ratio of welding time to

† The operating factor is sometimes referred to as the 'duty cycle', this is, however, liable to be confused with the duty cycle terminology which is used to describe the output rating of equipment.

non-welding time expressed as a percentage. Values of operating factor of 15% to 20% are not uncommon for MMA welding whilst figures of 30% to 50% may be achieved with manual GMAW [15,16]. Improvements in operating factor have a major influence on costs since they directly influence the labour element. The influence of operating factor on the labour cost is shown in figure 2.5.

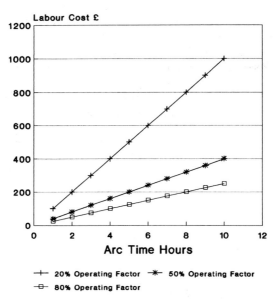

Figure 2.5   Effect of operating factor on costs (for a labour rate of £20 per hour).

(e)   *Post-weld operations*
The welding process may generate problems which need to be rectified after welding. The most common problems of this type are distortion and residual stresses, although metallurgical problems such as grain growth and hydrogen-induced cold cracking and cosmetic problems such as damage to surface coatings and spatter deposits must also be considered. The need to carry out additional mechanical or thermal operations after welding will obviously increase the cost of fabrication and process developments which reduce this requirement are desirable.

The risk of defects often generates a requirement for costly post-weld inspection and non-destructive examination, and although recent codes of practice allow the significance of defects to be related to the service conditions if rectification is required this involves increasing the value of work in progress, delays in production and is often labour intensive. Early identification of potential quality problems is therefore both desirable and cost effective.

2.2   Safety and Environmental Factors

Some of the potential safety hazards found in welding are listed in table 2.1 [17].

The operator is normally protected by means of protective clothing, local screening and ventilation whilst additional protection may be required to protect other workers in adjacent areas. These measures may be costly in themselves as well as having an effect on the overall efficiency of the production operation. Process developments which improve the working environment or remove the operator from the more hazardous operations are therefore desirable.

Table 2.1   Welding safety hazards.

Hazard	Process
Radiation, visible, infrared and ultraviolet.	MMA, TIG, MIG, plasma, laser
Ionising radiation (x-rays)	Electron beam
Particulate fume	Arc and power beam processes
Toxic gases (e.g. ozone)	Arc processes
Noise	Friction, plasma

2.3   Skill and Training Requirements

Many of the traditional welding processes required high levels of operator skill and dexterity, this can involve costly training programmes, particularly when the procedural requirements described above need to be met. The newer processes can offer some reduction in the overall skill requirement but this has unfortunately been replaced in some cases by more complex equipment and the time involved in establishing the process parameters has brought about a reduction in operating factor. Developments which seek to simplify the operation of the equipment will be described below but effective use of even the most advanced processes and equipment requires appropriate levels of operator and support staff training. The cost of this training will usually be recovered very quickly in improved productivity and quality.

2.4   Areas for Development

Advances in welding processes may be justified if they offer the following:

(i) increased deposition rate;
(ii) reduced cycle time;
(iii) improved process control;
(iv) reduced repair rates;
(v) reduced weld size;
(vi) reduced joint preparation time;
(vii) improved operating factor;
(viii) reduction in post-weld operations;
(ix) reduction in potential safety hazards;
(x) removal of the operator from hazardous area;
(xi) simplified equipment setting.

Some or all of these requirements have been met in many of the process developments which have occurred in the last ten years; these will be described in detail in the following chapters but the current trends in the application of this technology are examined below.

2.5   Process Application Trends

Several important trends may be identified on an international level in the application of welding processes; these are:

(i) process change in consumable electrode arc welding processes;

(ii) the increased use of automation;

(iii) increased interest in *new* processes (e.g. laser welding);

(iv) the requirement to fabricate advanced materials.

2.5.1 Consumable trends

The use of GMAW and flux cored arc welding (FCAW) processes at the expense of traditional MMA welding is evident in many industrialized countries. The figures for the production of welding consumables in the UK, Japan and the USA are shown in figure 2.6 [18]. It is expected that the amount of welding performed with MMA electrodes will stabilize at around 30% of all weld metal deposited in most industrialized countries whilst the use of submerged arc consumables has already stabilized at around

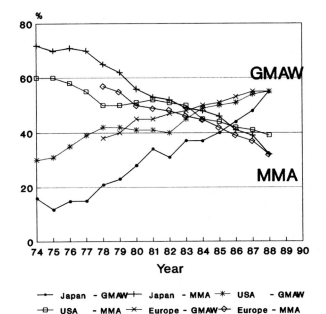

Figure 2.6 Consumable production trends 1970–88 (% of total output).

10% of deposited weld metal [19]. Significant increases in the use of flux-cored wires have already been observed in some countries (e.g. Japan) and it is expected that these trends will be repeated in the western world. The growth in cored wire sales is indicated in figure 2.7.

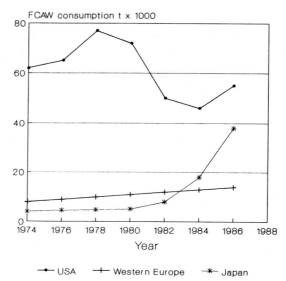

Figure 2.7   Growth in cored wire sales 1980–87.

This trend illustrates the importance of establishing the overall cost of the welding operation; the flux-cored consumable is inevitably more costly to manufacture and often more than four times more expensive to purchase when compared with the solid wires used for GMAW. The increase in deposition rate, higher operating factor, improved process tolerance and enhanced joint quality can, however, result in a reduction in the overall cost of the weld in spite of the higher consumable costs. The total cost will depend on the application; it can be shown for example that for a simple 6 mm horizontal/vertical fillet GMAW welding with a solid wire gives the lowest overall cost whereas in the case of a multipass vertical butt weld in 20 mm thick steel a rutile flux-cored wire is a more cost-effective solution [20].

2.5.2   Automation

The use of single-shot processes like resistance spot welding and continuous processes such as GMAW enable increased use to be made of automation. The trends for the introduction of robotics have been well reported and are summarized in figure 2.8 [21]. Many resistance welding robot applications are found in the automotive industry, whereas the use of robotic GMAW and GTAW systems is more diverse. It is also clear that the uptake of robots in Japan has exceeded that in western Europe and is also higher than the rate of application in the USA. The use of simple mechanisation and non-robotic automation in welding is less well reported, but in newer processes such as laser and electron beam welding some form of automation is an essential part of the system. Simple low-cost mechanization is recognized as a very cost-effective means of automation, particularly for GMAW and

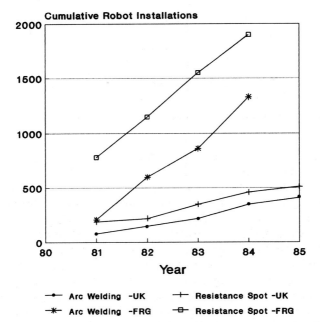

Figure 2.8   Trends in the use of welding robots in the UK and Germany (after Weston [21]).

FCAW processes, and its use is expected to increase. Computer-numerical-controlled (CNC) modular automation systems have been introduced in the last two years and these perform many of the functions normally associated with welding robots but allow increased flexibility.

The ability to integrate welding as a *well controlled* process in a flexible manufacturing facility is now technically feasible (figure 2.9), and with the aid of robot- or computer-controlled welding cells facilities such as on-line data recording, automatic component recognition, on-line quality assurance, and automatic reporting of

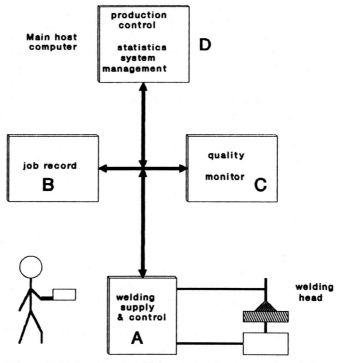

Figure 2.9 Integrated welding system. Operator selects job at A. Computer in the welding power supply sets parameters and supplies information. The job record is stored at B. On-line monitoring system C checks quality. Host computer D controls the system and provides a management report.

machine malfunction and production statistics may all be provided. The capital cost of integrated fabrication facilities of this type is high, but the economic benefits must be assessed on the basis of overall productivity improvements and end-product costs.

2.5.3 New processes

The use of new joining techniques such as laser welding, MIAB and diffusion bonding appears to be increasing. The application of these processes has in the past been restricted, but with the increased recognition of the benefits of automation and the requirement for high-integrity joints in newer materials it is envisaged that the use of these techniques will grow.

Total worldwide sales of industrial lasers have been growing at a rate of around 10% per year since 1988, when sales were estimated at around 3000 units [22], and it is expected that this growth rate will continue with CO_2 laser sales increasing at a rate of 13% per annum and 7% per annum growth in Nd:YAG. The YAG growth is expected to be dominated by high-power (1–2 kW) units. Welding applications are likely to be responsible for around 20% of this growth.

The viability of these processes has also been improved by equipment and process control developments, which will be discussed in the following chapters.

2.5.4 Advanced materials

There is an increasing demand to utilize more advanced materials as the service conditions of fabrications become more arduous and in many cases economic and environmental pressures favour improved strength to weight ratios. These trends are evident in such diverse areas as aircraft construction, offshore structures and the fabrication of microcircuits. Developments in advanced materials include higher-yield-strength thermomechanically treated low-alloy steels, fibre-strengthened composite materials (e.g. aluminium), polymers, cermets and ceramics. The application of these materials may depend on the ease of establishing reliable joints between two components of the same material or, quite often, between dissimilar interfaces, such as metal to ceramic bonds. Methods of bonding these advanced materials are still undergoing development, but procedures for high-yield-strength steels are already available using common arc welding techniques and encouraging results have been achieved using solid-phase bonding techniques for ceramic to metal joints.

2.6 Summary

The need for development in welding processes has been generated by economic and social factors, this has led to the development of more efficient consumables and equipment and a marked increase in the use of automation. A significant trend is the use of enhanced control and monitoring in conventional welding processes.

The introduction of new processes and advanced materials have, respectively, provided improved capabilities and renewed challenges in joining technology.

Examples of the advances made in some of the most important welding processes are described in the following chapters.

3 Welding Power Source Technology

3.1 Introduction

Many of the recent developments in arc welding have been made possible by improvements in the design of the welding power supplies and in particular the introduction of electronic control. The basic requirements of arc welding power supplies will be examined below and the principles of both conventional and advanced power source designs, their advantages and limitations will be described.

3.2 Basic Power Source Requirements

There are three basic requirements for arc welding power sources:

(i) to produce suitable output current and voltage characteristics for the process;
(ii) to allow the output to be regulated to suit specific applications;
(iii) to control the output level and sequence to suit the process and application requirements.

These requirements are illustrated in figure 3.1.

In order to produce suitable output levels for most arc welding processes the normal mains power supply must be converted from a high voltage–low current to relatively high current at a safer low voltage. This function may be performed by a conventional transformer, and if direct current is required a rectifier may be added to the output. The addition of a rectifier has the added advantage that a three-phase supply may be used† and the loading on the supply will be more uniform with approximately equal currents being drawn from each line. In addition the power source design must meet the following requirements:

(i) conformance with prescribed codes and standards;
(ii) safe installation and operation;
(iii) provide satisfactory operator interface/controls;
(iv) provide automation system interfaces, where necessary.

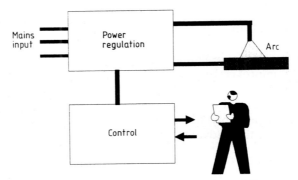

Figure 3.1 Functional requirements of welding power sources.

3.3　Conventional Power Source Designs

Conventional power sources have for many years used electro-magnetic control systems to enable the output power to be adjusted, some of the more common designs employ the following control techniques:

(i) tapped transformers;

† Although the 'Scott connection' system may be used to obtain a single-phase output from a three-phase supply this does not give balanced loading. Multi-operator transformers with a three-phase input and three single-phase outputs have been used but the loading of the three-phase input is dependent on the number of arcs in use and the output current.

(ii) moving iron control;
(iii) variable inductor;
(iv) magnetic amplifier.

3.3.1 Tapped transformer designs

By incorporating tappings in the primary coil of the welding transformer the turns ratio of the transformer may be varied and the
output regulated. It is normal to provide tappings which allow
adjustment to suit a range of mains input voltages in most
transformer-based designs, but when this technique is the principal
method of control additional tappings which are selected by a
switch as shown in figure 3.2 are provided. This type of control is
simple, robust and low in cost, but it will only provide a stepped
output and unless a large number of switch settings or dual-range
switching are provided the output voltage steps tend to be coarse.
Remote control or continuous regulation of the output are not
feasible with this system but it is often used for low-cost and light-
duty GMAW equipment. A typical switched output unit of this type
is shown in figure 3.3.

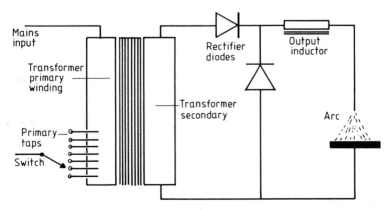

Figure 3.2 Tapped transformer/rectifier. Note that for clarity a single-
phase circuit is shown—most industrial units are three phase.

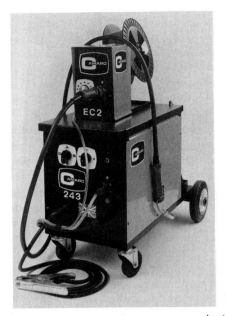

Figure 3.3   Typical tapped transformer GMAW unit. (Courtesy
of CAMARC.)

3.3.2   Moving-iron control

An alternative technique for modifying the output of a transformer is to vary the magnetic leakage flux with a shunt as shown
in figure 3.4. By controlling the position of the shunt the amount
of magnetic flux linking the primary and secondary coils is changed
and the output varies (the output varies inversely with the amount
of shunting). This method of control gives continuous variation of
the output and movement of the shunt may be motorized to allow
remote operation but it is costly, subject to mechanical wear and
the output can only be regulated slowly. The use of this type of
control is now largely confined to small low-cost MMA power
sources.

3.3.3   Variable inductor control

A variable or tapped inductor may be connected in the AC output
circuit of the transformer to regulate arc current as shown in figure

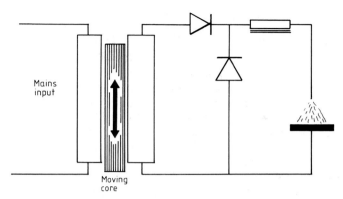

Figure 3.4   Moving-iron control transformer.

3.5. Although continuous current adjustment may be achieved with this design remote operation is not usually feasible and the high-current inductor is a large, costly item. A possible advantage of the design is that the inductance causes a phaseshift of up to $90°$ between AC current and voltage waveforms which may improve arc re-ignition (with $90°$ phaseshift the voltage will be at its maximum value when the current passes through zero). This design has been used in the past for MMA and GTAW power supplies.

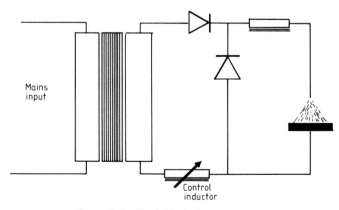

Figure 3.5   Variable inductor control.

3.3.4 Magnetic amplifier

A magnetic amplifier or saturable reactor control is illustrated in figure 3.6. A coil fed with a variable DC current is wound around a magnetic core which also carries a winding from the AC output of the transformer. As the DC level in the control coil is increased the average value of magnetic flux within the core increases towards the saturation level thus limiting the variation of the magnetic field and reducing the AC output. The technique allows continuous variation of the output, remote control and a certain amount of output waveform modification. The response rate of the system is, however, relatively slow, the DC control current can be fairly high (e.g. 10 A) and the saturable reactor is both bulky and expensive. Magnetic amplifier control has commonly been used for GTAW power sources although some GMAW equipment has also employed this technique in order to obtain some measure of remote control.

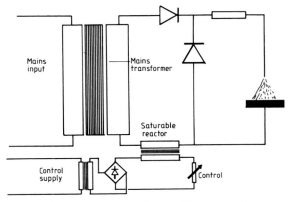

Figure 3.6 Magnetic amplifier control.

3.3.5 Control of static and dynamic characteristics of conventional power supplies

The dynamic (e.g. rate of change of current, and instantaneous relationship between current and voltage) and the static characteristics (the relationship between the mean output current and voltage) of the power supply can have a significant effect on process performance.

With conventional power sources it is normal practice to use constant-current static characteristics for the GTAW process in order to obtain optimum striking and current stability but constant voltage for GMAW in order to achieve self-adjustment [23]. These characteristics are normally predetermined at the design stage and cannot be varied by the user [24].†

Dynamic characteristics can be adjusted electrically, and in GMAW welding it is common to use a DC inductor in the power source output to control the rate of current rise during the short circuit in the dip transfer mode. The equipment shown in figure 3.3, for example, has a continuously variable inductance but many equipment manufacturers now limit the range of adjustment by using a simple low-cost tapped inductor.

The limited scope for adjustment of the dynamic characteristics by the user simplifies the operation of the equipment and offers adequate control for many applications, it does, however, restrict the possibility of significant improvements in process performance.

3.4 Electronic Power Regulation Systems

The availability of high-power semiconductors has led to the development of a range of alternative, electronic power source designs [25] which may be classified as follows:

SCR phase control;
transistor series regulator;
secondary switched transistor;
primary rectifier–inverter;
hybrid designs.

3.4.1 SCR phase control

Silicon-controlled rectifiers (SCRs) may be regarded as switchable diodes. The device only starts to conduct in the forward direction

† A conventional variable slope/variable inductance unit has been developed but this is costly and complex [23].

when a signal is applied to the gate connection. Under normal circumstances the device cannot be turned off until the forward current falls to zero. These devices may be used instead of the normal diodes in the secondary circuit of a DC power supply. To regulate voltage output the delay between the normal onset of conduction and the gate signal is varied (figure 3.7). If the amplitude of the voltage waveform is fixed it is necessary to use a long firing delay to achieve low output levels and the ripple in the output waveform becomes severe. This problem may be reduced by using a three-phase SCR bridge, by using a large output inductance, or by using an interphase inductance. Alternatively the SCR control may be placed in the primary of the transformer in which case some smoothing is obtained from the transformer itself. The use of inductance in any of the forms described above is an effective means of smoothing, but it does limit the dynamic response of the power source.

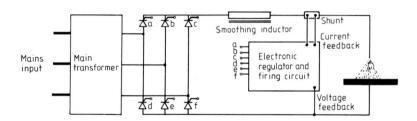

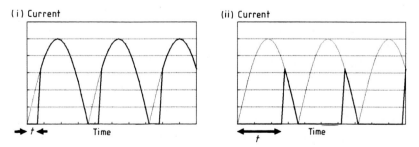

Figure 3.7 SCR phase control circuit (three phase) and controlled wave form for a single phase. a–f in the upper diagram are SCRs. The lower traces show the waveform from a single SCR: (i) small firing delay with high mean output; (ii) large firing delay with low mean output.

An AC output may be obtained by using SCRs connected 'back to back', one set conducting in the positive half-cycle whilst another set conducts in the negative half-cycle. In addition, by using an inductor or an inverter circuit (figure 3.8) it is possible to produce a 'square' output waveform which offers process benefits in GTAW, MMA and SAW.

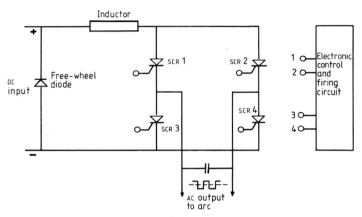

Figure 3.8 Simple secondary inverter for AC welding.

The advantages of this type of control are its simplicity, robustness and the large amplification obtained which enables high output levels to be controlled by very low-level electronic signals. The speed of response of the system is limited by the necessity to cross current zero before a revised firing angle becomes effective; hence the best response which could be expected would be in the range of 3 to 10 ms.

Even with the limitations of ripple and response rate it is possible to produce power sources with significantly better performance than previous conventional designs, and in particular it is possible to stabilize the output of the power source by means of feedback control.

The SCR phase control system has been used for DC MMA, GMAW, square-wave AC GTAW [26,27], GMAW and SAW power sources.

3.4.2   Transistor series regulator

The output of a transistor may be controlled by adjusting the small current flowing through its 'base' connection. The series regulator consists of a transistor in series with the DC welding supply, the output power being continuously regulated by means of the base current. It is usual to incorporate a feed-back control system to ensure output stabilization and an amplifier to supply the drive signal to the transistor (figure 3.9).

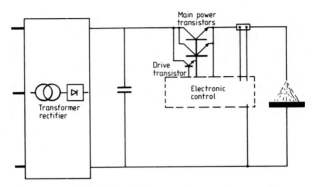

Figure 3.9   Principle of the transistor series regulator.

Until fairly recently the capacity of individual transistors was limited and large banks of devices (connected in parallel) were necessary to handle typical welding currents (figure 3.10). The recent availability of higher-power devices has, however, reduced this problem.

The important characteristics of the transistor series regulator are its rapid response rate (the response time of transistors is measured in microseconds) and its 'ripple'-free output [28]. The main disadvantage of the system is the poor efficiency and high cost. The poor efficiency results from the dissipation of surplus power as heat from the devices, this necessitates the use of water cooling for most applications. The equipment cost is a function of the number of devices used and the need to balance each transistor to ensure current sharing. The high response rate, accuracy and low ripple of this type of power supply make it suitable for small high-precision supplies and particularly for process research work. GTAW, GMAW and SAW units are available.

Figure 3.10 Typical 500 A series regulator. The upper half shows the power transistors on water-cooled aluminium heat sinks and the lower half the smoothing capacitors.

3.4.3 Secondary switched transistor

The high heat dissipation of the transistors in a series regulator results from their continuous operating mode. An alternative method of regulating the output is to switch the transistor on and off at a rapid rate; the mean output level is then a function of the ratio of on to off time (figure 3.11).

Although the circuit design is very similar to that of the series regulator (figure 3.12) the process of switching the device in this way gives a significant improvement in efficiency and enables normal air cooling to be used. The chopped waveform appears at the output, but if a sufficiently high switching rate is used this does not have any detrimental effect on the process. Switching frequencies of 1000 to 25 000 Hz are commonly used.

Response rate is also determined by the switching frequency but the higher-frequency supplies are capable of responding within a few microseconds, which is significantly faster than conventional supplies and approaching the rate achieved with series regulators.

GMAW and GTAW power sources of this type are available and offer high precision at currents up to 500 A at an economical capital cost.

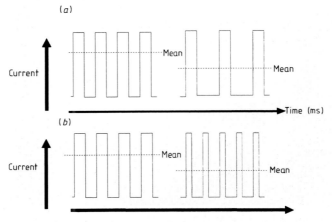

Figure 3.11 Modulation techniques for control of output level. (*a*) Frequency modulation. (*b*) Pulsewidth control.

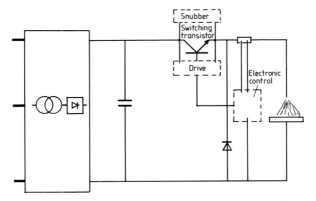

Figure 3.12 Circuit principle—switched control.

3.4.4 Primary rectifier–inverters

The methods of control outlined above use a conventional transformer to achieve the step down in voltage required for welding.

This transformer operates at the incoming mains frequency of 50 Hz.

The primary inverter design utilizes the fact that transformer size may be significantly reduced if its operating frequency is increased as shown in figure 3.13. The basic circuit is illustrated in figure 3.14, and the principle of operation is illustrated in figure 3.15.

Figure 3.13 200 A welding power source transformers: left, 20 kHz inverter (rectified and inverted three phase); right, conventional three phase.

The primary AC supply is first rectified and the resultant high DC voltage is electronically converted by the inverter to high-frequency AC. Only at this stage does the supply enter the transformer. Since the frequency of operation is between 5000 and 50 000 Hz the transformer is small, furthermore output control is achieved by chopping or phaseshifting within the inverter and very high response rates are achieved. The transformer output must be rectified to avoid potential losses in the high-frequency AC circuit.

The welding output is smoothed and stabilized, and although it is not possible to achieve the same response rates as those obtained with the series regulator, it is possible to produce the output characteristics required for recent process control developments.

This type of circuit was initially used for MMA power sources but it is now being employed for GTAW and pulsed GMAW units. It has particularly good electrical efficiency and a comparison of inverter and conventional power sources at current settings of 250 A has shown that idle power consumption is only one-tenth of that of a conventional machine and during welding the efficiency is around 86% compared with 52% for a conventional unit [29].

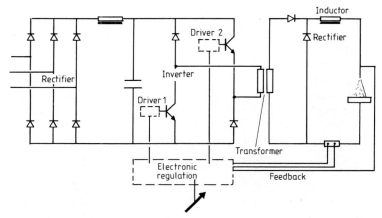

Figure 3.14　Typical inverter circuit.

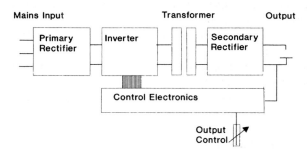

Figure 3.15　Primary rectifier–inverter.

3.4.5　Hybrid designs

It is possible to combine the electronic control techniques outlined above to improve the performance and cost effectiveness of the

power source. For example, the use of a secondary chopper to pre-
regulate the supply followed by a small air-cooled transistor series
regulator for final control of the output has been described [30],
the circuit is shown schematically in figure 3.16 and the advantages
of this approach are summarized in table 3.1. Hybrid designs may
also be adopted to produce a square-wave AC output by adding a
secondary inverter to the output of a DC phase-controlled unit.
SCR phase-controlled power sources may be used in conjunction
with an SCR inverter [31] or alternatively the system may be based
on an integrated primary rectifier/inverter design. Sophisticated
experimental hybrid units have been developed, for example for
high-frequency AC plasma welding [32] in order to investigate
potential improvements in process control.

Alternative power devices such as asymmetrical SCRs (ASCRs)
[33] or metal oxide–silicon field-effect transistors (MOSFETs) may
also be used to improve the efficiency of conventional electronic
and hybrid systems.

3.4.6   Features of electronic power source designs

The electronic designs all share a capability for remote control and
are easily interfaced with system controllers within the power
supply or from an external source. The output response, accuracy
and repeatability are generally considerably better than those
achieved with conventional electromagnetic control systems but
the features of the various designs are summarized in table 3.2.

It is not possible to select an ideal design from this list but series
regulator designs are often only justified for very-high-precision
and research applications whilst primary inverter-based designs are
cost effective and suitable for a wide range of production tasks.
One important characteristic which all of these systems share is the
use of feedback control.

(a)   *Feedback control*

Feedback control is a useful technique which can be applied most
effectively with electronic power sources. The basis of the technique is illustrated in figure 3.17; the output of the system is
measured and compared with the desired output parameters, any
difference between the two values will cause an 'error' signal to be
generated. This signal is fed back to the power control system

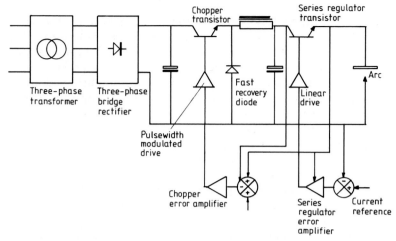

Figure 3.16 Hybrid chopper/series regulator power source for GTAW (after Rodrigues [30], courtesy PEC Ltd).

Table 3.1 Comparative evaluation of 300 A power sources.

Feature	Hybrid chopper series regulator	Conventional series regulator
Maximum dissipation	2 kW	8.5 kW
Devices (transistors)	Four high current	> 100 low current
Ripple	< 2 A peak/peak	< 1 A peak/peak
Overall cost	Cheaper to make	More costly (labour)
Response	Max 150 Hz, 1 ms rise-time	Max 1 Hz, 100 μs rise-time
Accuracy	Better than 1% of maximum current	Better than 1% of maximum current
Chopping frequency	1.5 Hz (evaluation unit)	—

Table 3.2 Features of various electronic power source designs.

Power source type	Output characteristics	Electrical efficiency	Physical characteristics	Relative cost	Applications
Conventional tapped transformer rectifier, moving-iron, variable inductor, magnetic amplifier, etc	Fixed at design stage, slow response, no mains stabilization	Fair	Large, heavy, robust, reliable	1	Manual SMAW GTAW. General-purpose fabrication
SCR phase control	Electronically variable within response limits of system. Mains stabilized. High ripple	Fair	More compact than equivalent conventional design	3	Manual and mechanized GTAW/GMAW and manual SMAW. Medium to high quality
Transistor series regulator	Very fast response, flexible control, accurate, ripple free, repeatable	Poor	Fairly large; may be water cooled	6	High-quality GTAW/GMAW. Pulsed output R&D, precision
Primary rectifier inverter	Fast response, variable output, stable and repeatable	Very good	Compact electronically complex	4	Medium to high quality, multi-process
Hybrid and secondary chopper	Fast response, variable output, stable and repeatable	Very good	Medium size, air cooled	4	Medium to high quality, multi-process

which adjusts the output to correct the imbalance. Although this type of control may be applied to conventional power source designs it is usually costly, complicated and too slow, hence most conventional power sources have 'open-loop' control, i.e. if the input varies the output changes by a proportional amount. The higher response rates and low signal levels available from electronic control systems make 'closed-loop' or feedback control effective and economical and give inbuilt stabilization of output.

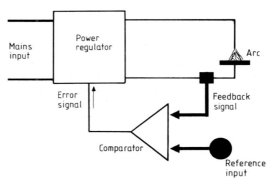

Figure 3.17 Principle of feedback control.

3.5 Output Level, Sequence and Function Control

For most welding processes it is required to follow a prescribed sequence when initiating or terminating the process. In addition it may be necessary to control the rate of current rise or decay and modulate the output during welding. The extent to which these features could be provided has in the past been limited by conventional power source design technology and the use of relay logic and electrical control techniques. Two new approaches have now been adopted in order to improve the flexibility and accuracy of sequence control, these are:

(i) the use of discrete electronic control;
(ii) microprocessor control.

3.5.1 Discrete component electronic control

The control signal levels required to 'drive' the electronic power regulation circuits described above are usually small and may be derived from electronic logic circuits. These circuits may be configured to perform the most complex tasks using standard analogue and digital components (such as timers, programmable logic arrays, power regulators, operational amplifiers and comparator circuits) which are packaged in single chips. The performance of these systems is far better in terms of cost, speed, accuracy and long-term reliability than the previous relay logic designs. In general, however, discrete electronic control circuits are custom designed for a specific power source and the facilities and range of operation are fixed at the design stage. The flexibility of discrete electronic circuit designs has been increased by the storage of welding control parameters on electrically programmable read only memory (EPROM) chips which may be easily programmed by the manufacturer of the equipment and replaced when improved process parameters are developed or new facilities are added.

3.5.2 Microprocessor control

The alternative approach, using microprocessor control, can allow much more flexibility and many additional facilities may be provided. A single microprocessor chip can control both the sequence of welding and regulation of the output power.

The schematic design for a microprocessor control system is illustrated in figure 3.18. The microprocessor carries out a series of instructions and calculations sequentially but certain important tasks may be given priority over less important tasks, for example it may be required to check the output current level once every 0.3 ms when welding is in progress but the status of certain front-panel controls may be ignored unless welding ceases. The effectiveness of this type of system for real-time control of the output depends on the resolution of the analogue to digital converters, the operating speed of the microprocessor (the clock rate) and the design of the software. A typical system using a clock rate of 12 MHz and 10-bit analogue to digital converters is able to check and correct any deviations in output every 0.3 ms and maintain the current within 1% of the desired level.

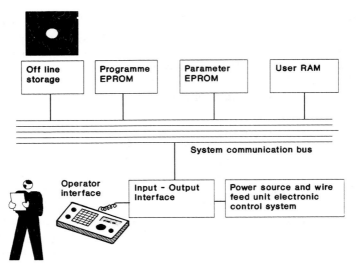

Figure 3.18   Microprocessor control system.

In practice many of the individual components shown in figure 3.18 may be combined in a single microcontroller chip, and a typical power source control board using an Intel 8097 controller for a sophisticated multiprocess power source is shown in figure 3.19. Although the dedicated microprocessor control approach

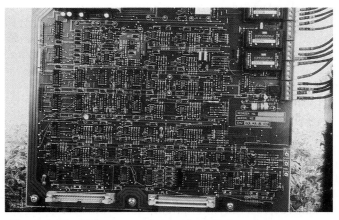

Figure 3.19   Main control board. 500 A microprocessor-controlled primary rectifer–inverter power source.

does not allow complete design flexibility (the software will often represent a significant investment and revisions may be costly) the designer can build in the ability to change certain parameters based on a knowledge of the welding process requirements. This may be used to simplify the operation of the equipment or to provide the user with the ability to reprogramme key process variables.

3.5.3 Programming and one-knob control

The concept of a single adjustment knob for 'complex' parameter setting, for example in the GMAW process, is not new and power sources using preset wire feed and voltage controls and a single condition selector switch were available in the mid 1970s [33]. However, without electronic feedback control novel but unreliable methods of mains voltage stabilization had to be employed. With electronic power regulation and feedback control power sources may be programmed by the supplier or the user with reliable 'optimum' welding conditions. Programming and storage of welding

Figure 3.20 Computer-controlled underwater welding power supply: left, high-voltage 300 A transistor series regulator; right, operator interface—IBM 286 computer (linked by RS232 to control computer—not shown).

parameters is made even easier if microprocessor control systems are used as described above.

3.5.4   Computer control

Control of electronic power sources using an external micro-computer has also been used; this has mainly been for research applications where a wide range of process variables are under investigation but many microprocessor-controlled power sources now have the facility to communicate with a host computer using standard serial communications protocols (RS232, RS423, etc). This allows welding parameters to be 'downloaded' to the equip-ment as well as facilitating remote control and monitoring. This technique may also be used in production applications and the system for remote-controlled welding of underwater pipelines illus-trated in figure 3.20 is a good example of this approach.

3.6   Practical Implications of Electronic Power Regulation and Control

The changes in the technology of welding power sources described above have some significant practical implications: the power sources can be manufactured using modern electronic assembly techniques and the dependence of these designs on expensive raw materials, such as iron for transformer cores and copper for the windings, is reduced. This should enable the manufacturers of these more advanced power sources to offer them at costs compar-able with those of conventional designs. These designs also offer the user the following advantages:

(i)   improved repeatability;
(ii) increased ease of setting;
(iii) enhanced process capabilities.

Improved repeatability has a direct impact on the quality of the welded joint, the ability to maintain welding parameters within the range specified in the welding procedure, and is likely to reduce the repair and rework costs discussed in Chapter 2. The increased ease of setting should improve the operating efficiency and reduce the risk of operator error. The enhanced process capabilities result from the ability to change various process output parameters of an electronic power supply during welding. The output characteristics are not predetermined and may be varied (within the limits of the transformer output) to produce beneficial effects. For example, in the case of GMAW constant-current output characteristics may be used for improved control and the output may be modified dynamically to provide self-adjustment. In MMA welding systems the current may be increased instantaneously at low voltages to prevent electrode 'sticking'. These characteristics will be discussed in more detail in the following chapters.

In order to make effective use of electronic power supplies the user will, however, need to consider service support and training. The skills required for repairing and maintaining this type of equipment are not the same as those needed for conventional electromagnetic power sources.

Figure 3.21 Typical multiprocess, microprocessor-controlled inverter power source (Commander BDH 320, courtesy of Migatronic).

3.7 Summary

Conventional power source designs continue to be viable for simple, robust low-cost applications. A tapped transformer/rectifier GMAW machine is, for example, approximately 60% cheaper than an electronic unit of the same current rating. However, the electronic power source designs usually offer improved capabilities in terms of output consistency and flexibility as well as providing the possibility of enhanced process control [34].

Advanced welding power supplies use a range of electronic power regulation techniques, and either discrete component electronic circuitry or microprocessor control of the operating sequence and output level. A typical microprocessor-controlled multiprocess inverter power source of this type is shown in figure 3.21.

4 Filler Materials For Arc Welding

4.1 Introduction

Some of the more important consumables used in the fusion welding processes are:

(i) coated electrodes for MMA welding;
(ii) wires and fluxes for SAW;
(iii) filler wires for GMAW and FCAW.

The design and formulation of consumables for the traditional welding processes such as MMA and SAW have continued to evolve largely in response to the need to match improvements in the properties of new materials but also to enhance the operating tolerance and stability of the processes. Filler wires for the continuous-feed consumable electrode processes such as GMAW have changed little but significant progress has been made in the development and application of flux-cored wires.

4.2 MMA consumables

An extensive range of MMA consumables is now available covering the joining requirements of the more important engineering materials as well as repair, surfacing, cutting and gouging electrodes.† The main developments which have taken place in these

† A summary of the latest UK and US classifications for carbon steel electrodes will be found in Appendix 2.

consumables have been in the following areas:

(i) improved toughness;
(ii) improved hydrogen-controlled electrodes for ferritic steel;
(iii) improved performance stainless steel consumables.

MMA has been used extensively in the shipbuilding, offshore, and power-generation industries for the fabrication of carbon–manganese and low-alloy steels. The achievement of good toughness and resistance to hydrogen-induced cold cracking are important considerations in these applications.

4.2.1 Improved toughness

Many structures, particularly in the offshore and cryogenic industries, are expected to operate at temperatures well below 0 °C. Considerable improvements in weld metal toughness of ferritic

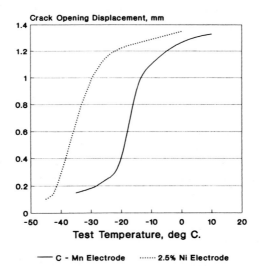

Figure 4.1 Improvements in notch toughness for MMA consumables (after Dawson G W and Judson P 1982 Procedural guidelines for the achievement of tough welded joints in structural steels for offshore applications *Proc. 2nd. Int. TWI Conf. 'Offshore Welded Structures'* (Cambridge: The Welding Institute) paper 2, pp. 1–13).

materials at temperatures below $-40\,^\circ$C have been obtained by control of electrode formulation and the addition of nickel [35]. The effect of this improvement on the notch toughness of a series of ferritic MMA electrodes is shown in figure 4.1.

4.2.2 Improved hydrogen control

Hydrogen-induced cold cracking has been a significant problem with low-alloy and higher-carbon steels particularly when thicker sections are welded. Control of the hydrogen content of the weld metal may be used to avoid this problem and this control may in turn be improved by electrode formulation, storage and packing [36]. Particular attention has been given to limiting the reabsorption of moisture (the main source of hydrogen) by the electrode coating, this is achieved by careful selection of the coating constituents and special packaging. Basic hydrogen-controlled electrodes which will give weld metal hydrogen contents of less than 5 ml/100 g with ambient conditions of $35\,^\circ$C and 90% humidity for up to ten hours after opening the packaging are now available [37].

4.2.3 MMA electrodes for stainless steel

The operating performance of common austenitic stainless steel MMA electrodes has been considerably improved by the introduction of rutile (TiO_2) based flux coatings. These coatings give improved arc stability and excellent weld bead surface finish. Electrodes have also been developed for the fabrication of the new corrosion-resistant alloys and in particular duplex and high-molybdenum stainless steels [38].

4.3 Submerged Arc Welding Consumables

The submerged arc process is well established and a standard range of wires and fluxes has been devised for the most common applications. Two recent advances in the process are:

(i) the development of high-toughness consumables;
(ii) the use of iron powder additions.

4.3.1   High-toughness consumables

Higher-toughness consumables have been developed in response to the requirement for reasonable impact properties down to $-40\,^{\circ}\mathrm{C}$ for offshore structures. This has been achieved by the use of wires which are microalloyed with titanium and boron and a semi-basic flux [39]. Typical Charpy-V notch curves showing the improvement in toughness compared with a conventional SD3 molybdenum wire are shown in figure 4.2.

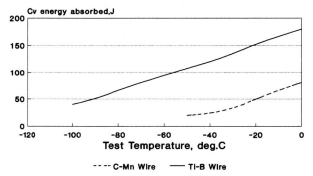

Figure 4.2   High-toughness SAW consumables: lower bound Charpy-V impact transition curves for different SAW wires (after Taylor D S and Thornton C E 1989 *Welding Review* **8**(3)).

4.3.2   Addition of iron powder

The addition of iron powder to the submerged arc weld increases the deposition rate of the process by more than 60% [40–42] as well as offering improvements in weld metal quality. The technique takes advantage of the fact that excess arc energy is normally available in submerged arc welding; this usually results in increased melting of the parent plate and high levels of dilution. If a metal powder is added to the weld pool some of the arc energy is dissipated in melting this powder and the additional metal which results improves the joint completion rate. Unlike other methods of increasing the productivity of the process this technique reduces the heat input.

Iron powder may be added to the joint in three ways:

(i) prefilling the joint prior to welding;
(ii) using a forward-feed system;
(iii) magnetic feeding on the wire surface.

It has been found [43] using the prefilling technique that a considerable reduction in overall heat input may be obtained as well as a significant improvement in productivity. This could be advantageous for the new fine-grained, high-strength low-alloy steels since it limits the adverse effects of high heat inputs on the thermo-mechanical treatment which is used to obtain the improved properties.

The technique has been applied to the fabrication of high-toughness steels for offshore applications and using the forward-feed system, a basic flux and a 1.7% Mn wire Charpy-V notch values of 173 J at $-40\,^{\circ}C$ have been reported [39] for a 50 mm thick single-V butt weld in BS 4360 50DD material.

4.4   Filler Wires for GMAW and FCAW

Increased utilization of the GMAW process has prompted some development of solid wire consumable/shielding gas packages but the most significant developments in this area are related to flux-cored consumables [44].

4.4.1   Solid filler wires for GMAW

Solid filler wires usually have a composition which is nominally the same as the material being joined. Minor chemical changes, for example in deoxidants, have been shown to enhance transfer and bead shape but scope for change in this area is limited. Although early attempts were made to improve metal transfer by surface treatment [45] this is not normally practicable due to the likelihood of surface damage.

Recent work on rare earth (cerium) additions to the filler wire indicates that transfer of steel wires in pure argon can be improved, but this is of little practical significance unless very low levels of weld metal oxygen are being sought, (e.g. in the welding of 9% Ni steel with matching fillers). Oxidation of the metal as it transfers across the arc may reduce the level of certain alloying elements and this will in turn depend on the oxidation potential of the gas and the reactivity of the element involved.

It has also been found [46] that very small changes in the residual chemical composition of ferritic steel wires can have a marked effect on low-temperature toughness. These changes are insufficient to affect the weld metal property requirements for most applications but may be significant in critical low-temperature and cryogenic joints.

In carbon steel wires it has been found that excessively thick copper coating can cause feedability problems, and it was feared at one time that copper fumes from the coating may represent a health hazard. As a result uncoppered wires became available. These wires give good feedability, although tip wear is a potential problem; however, it is now common for copper coating thickness to be controlled and it has been demonstrated that this overcomes fume and feedability problems.

4.4.2 Flux-cored wire

Flux-cored wires consist of a metal outer sheath filled with a combination of mineral flux and metal powders (figure 4.3). The FCAW process is operated in a similar manner to GMAW welding and the principle is illustrated in figure 4.4. The most common production technique used to produce the wire involves folding a thin metal strip into a U shape, filling it with the flux constituents, closing the U to form a circular section and reducing the diameter of the tube by drawing or rolling.

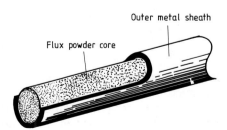

Outer metal sheath

Flux powder core

Figure 4.3 Construction of a flux-cored wire.

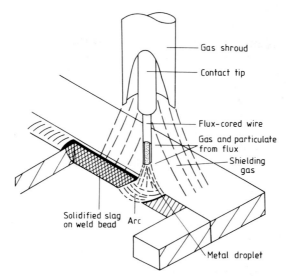

Figure 4.4   Principle of operation of FCAW.

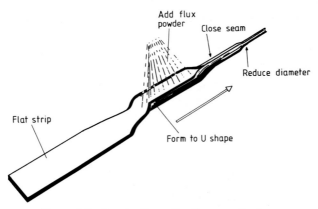

Figure 4.5   Production of a flux-cored wire.

The process is illustrated diagrammatically in figure 4.5. The seam is closed during the reduction process. Alternative configurations (figure 4.6) may be produced by lapping or folding the strip or the consumable may be made by filling a tube with flux followed by a drawing operation to reduce the diameter. Typical finished wire diameters range from 3.2 to 0.8 mm. Flux-cored wires offer the following advantages

(i) high deposition rates;
(ii) alloying addition from the flux core;
(iii) slag shielding and support;
(iv) improved arc stabilization and shielding.

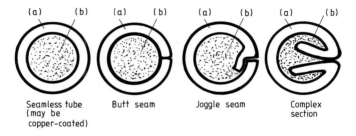

Figure 4.6 Alternative configurations for flux-cored wires.

(a) *Deposition rate*

The deposition rate will be substantially higher than that normally achieved with MMAW and marginally better than that obtained with a solid wire GMAW. This increase in deposition rate is attributable to the increased current density and the fact that all of the current is carried by the sheath. The deposition rate will, however, depend on the thickness and resistivity of the sheath material, the polarity and the electrode stick-out (figure 4.7). The melting rate MR of a flux cored wire may be expressed as:

$$\text{MR} = k + \alpha I + \frac{\beta l I^2}{A} \qquad (4.1)$$

where K, α and β are constants, I is the mean current, l is the stick-out length and A is the cross sectional area of the conductor. The term αI represents arc melting whilst the $\beta l I^2/A$ term indicates the resistive heating in the wire extension. †

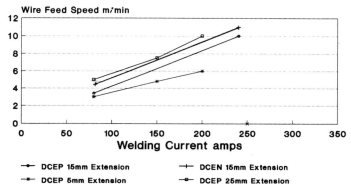

Figure 4.7 Effect of polarity and electrical extension on the burn-off rate of a basic flux-cored wire.

It is obvious from this equation that significant increases in burn-off rate may be achieved by increasing the wire extension and the operating current.

(b) *Alloying addition*
The range of compositions of solid GMAW wires is limited by the technical and commercial difficulties involved in producing relatively small quantities of special compositions. Flux-cored wires can, however, be modified by minor adjustments in the flux formulation to produce a range of weld metal compositions and operating characteristics. The range of compositions currently available for plain carbon and alloy steels‡ is similar to that for MMA electrodes; with rutile (TiO_2) based formulations for ease of

† Typical burn-off curves which illustrate this relationship will be found in Appendix 3.
‡ The UK and US specifications for flux-cored welding consumables are summarized in Appendix 4.

operation, basic (CaO) high-toughness, hydrogen-controlled formulations and metal powder cores for high recovery and low slag formation.

It is also possible to extend this technique to produce low-cost austenitic stainless steel or highly alloyed hardfacing deposits from wires with a plain carbon steel sheath.

(c) Slag shielding and support
The solidification characteristics of the slag may be designed to enhance the performance of the process. For example a fast freezing Rutile slag may be used to support the weld pool in vertical or overhead welding enabling higher operating currents, improved productivity and better fusion characteristics to be obtained. Alternatively the slag characteristics may be adjusted to provide additional shielding and control of bead shape. This is particularly important in the case of the stainless steel consumables discussed below.

(d) Arc stabilization and shielding
The decomposition of the flux constituents may be used to generate shielding gases as in MMA welding, for example CO_2 may be produced by the decomposition of calcium carbonate

$$CaCO_3 \rightarrow CaO + CO_2. \qquad (4.2)$$

Arc ionizers may also be added to the flux to obtain improved running characteristics and arc stability. It is possible using these techniques to produce electrodes which operate with alternating current or DC electrode negative and this may have beneficial effects on the melting rate and weld bead properties.

4.4.3 Modes of operation

Flux-cored wires may be operated successfully with or without an additional gas shield.

(a) Self-shielded operation
In self-shielded flux-cored wires the flux must provide sufficient shielding to protect the molten metal droplets from atmospheric contamination as they form and transfer across the arc. Since some nitrogen and oxygen pick up is inevitable the weld metal chemistry

is often modified to cope with this (by adding aluminium for example). In addition to this shielding action the flux must also perform arc stabilization, alloy addition and slag control functions. Formulation of suitable flux compositions is consequently difficult but several successful consumable designs are available. These self-shielded wires do have benefits for site use where moderate side winds are experienced but the demands on the design of flux may be reflected in poor process tolerances; for example for some positional structural wires the operating voltage range must be held within ± 1 V of the recommended level to produce the required mechanical properties and prevent porosity.

These requirements are reduced if an additional shielding gas is used.

(b) *Gas-shielded operation*
Auxiliary shielding may be provided if a conventional GMAW torch is used. For steel it is common to use either CO_2 or argon/CO_2 mixtures for this purpose, this allows the positional performance, mechanical properties and process tolerance to be improved and in spite of the additional cost of the shielding gas the overall cost of the process may often be reduced.

4.4.4 Types of flux-cored consumable
The following groups of flux-cored wires have been developed:

  (i) plain carbon and alloy steel;
  (ii) hardfacing alloys;
  (iii) stainless steel.

(a) *Plain carbon and alloy steels*
Details of some of these wires are listed in Appendix 5 but for discussion they may be subdivided into:

  (i) rutile gas-shielded;
  (ii) basic gas-shielded;
  (iii) metal cored;
  (iv) self-shielded.

(b) Rutile gas-shielded

Rutile gas-shielded wires have extremely good running performance, excellent positional welding capabilities, good slag removal
and provide mechanical properties equivalent to or better than
those obtained with a plain carbon steel solid wire. By alloying
with nickel good low-temperature toughness (e.g. 100 J at
$-40\,°C$) may be achieved.

(c) Basic gas-shielded

Basic gas-shielded wires give reasonable operating performance,
excellent tolerance to operating parameters and very good mechanical properties. Alloyed formulations for welding low-alloy and
high-strength low-alloy steels are available. The positional performance of these wires, particularly in the larger diameters, is not
as good as that of the rutile consumables.

(d) Metal-cored–gas-shielded

Metal-cored wires contain very little mineral flux, the major core
constituent is iron powder or a mixture of iron powder and ferro
alloys. These wires give very smooth spray transfer in argon/CO_2
gas mixtures, particularly at currents around 300 A although they
may also be used in the dip and pulse modes (see Chapter 6) at low
mean currents. They generate minimal slag and are suitable for
mechanized applications.

(e) Self-shielded

Self-shielded wires are available for general-purpose downhand
welding, positional welding and a limited range of wires are
available for applications which require higher toughness. As in
the rutile wires the higher toughness requirements are usually met
by alloying with nickel. Considerable use has been made of these
consumables in offshore applications [48] where it has been demonstrated that by close control of the operating parameters consistently high toughness values may be achieved under site conditions.

(f) Hardfacing and surfacing alloys

A wide range of hardfacing and surfacing alloys are produced in
the form of flux-cored wires.† These include plain carbon steels,

---

† A summary of the normal range of hardfacing consumables is provided
in Appendix 5.

austenitic stainless steels, alloys containing high chromium and tungsten carbide and nickel- and cobalt-based consumables. Many of these wires are self-shielded and intended primarily for site use. The running performance is not normally as good as that found in the constructional wires described above due to the increase in the ratio of alloying elements to arc stabilizers in the core material but they provide a cost-effective means of depositing wear- and corrosion-resistant material.

(g) Stainless steel

Stainless steel flux-cored wires have recently been introduced and matching consumables are available for most of the common corrosion-resistant materials. Both gas-shielded metal-cored and rutile-based formulations are available with the latter giving exceptionally good operating characteristics, wide process tolerance, low spatter and excellent surface finish (figure 4.8).

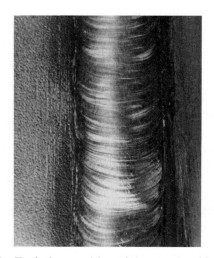

Figure 4.8 Typical austenitic stainless steel weld made with the FCAW process. The photograph shows the excellent weld bead surface finish of a butt weld in 3 mm thick 304L stainless steel plate made in the vertical up position (3G) using a rutile flux-cored wire and argon/20% CO_2 shielding.

4.4.5   Practical considerations

Gas-shielded flux-cored wires are often easier to use than the solid wire GMAW process but certain differences exist in operating technique. The sensitivity of these consumables to stick-out has been indicated above. Long electrode extensions give higher burn-off rates but the permissible extension may be limited in the case of gas-shielded wires by loss of effective secondary shielding. In the case of self-shielded wires where exceptionally long wire extension may be required to achieve high deposition rates an insulated guide incorporating fume extraction may be recommended (figure 4.9).

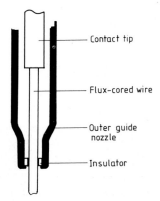

Figure 4.9   Insulated extended electrode guide and fume extraction system.

Very short extensions may be undesirable, for example it has been found that with rutile wires designed for positional use the combination of fast-freezing slag and excess surface lubricant on the wire can cause surface porosity if short electrode extensions are used, particularly in the downhand position. Increasing the extension allows the excess lubricant to be driven off and avoids the problem. For manual operation the minimum extension may be controlled by the relative position of the gas shroud and contact tip as shown in figure 4.10.

The equipment required for flux-cored wire operation is basically the same as that required for GMAW welding although for the less tolerant types of self-shielded consumable a voltage-stabilized power source may be specified (any of the electronic designs discussed in Chapter 3 will meet this requirement). For all flux-cored wires it is important to use specially designed feed rolls to avoid crushing the wire in the feed system.

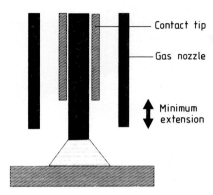

Contact tip

Gas nozzle

Minimum extension

Figure 4.10 Limiting the minimum electrode extension by means of contact tip to gas nozzle adjustment.

4.4.6 Applications of FCAW

The applications of flux-cored wires include the joining of thick-section high-strength steels for critical applications as illustrated in figure 4.11, high-speed mechanized welding of lighter sections using metal-cored wire or the fabrication of high-quality stainless steel process plant.

4.4.7 Limitations of flux-cored wires

The apparent limitations of flux-cored wire are:

   (i) cost;
   (ii) fume;
   (iii) consistency of the consumable.

(a) Cost

The cost of flux-cored wires may be four times that of a solid wire but this must be considered in the light of potential improvements

Figure 4.11 Dragline bucket fabricated from quenched and tempered steel using a basic flux-cored wire.

in productivity and the fact that the cost of consumable represents a relatively small part of the total fabrication cost (see Chapter 2). In fact the use of a flux-cored wire can often reduce the total cost; for example in trials on a vertical V butt joint in 25 mm thick BS 4360 50D material [49] it was found that the use of a rutile flux-cored wire enabled a saving of 28% in the cost of the joint compared with GMAW welding with a solid wire. The saving resulted from a decrease in labour costs due to the increased welding speed (the flux-cored wire could be used at a higher mean current). The results of these tests are shown in figure 4.12.

(b) *Fume*
Due to the high burn-off rate, the presence of mineral flux constituents and the continuous mode of operation it is inevitable that flux-cored wires will produce more particulate fume than either MMA or GMAW welding with a solid wire. Whilst most of this particulate may be considered to be fairly inert dust some of the consumables and flux constituents give rise to substances which are thought to be toxic; hexavalent chromium from chromium-bearing consumables and barium compounds found in the fume of some self-shielded wires are the main areas of concern.

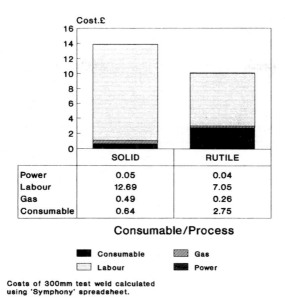

	SOLID	RUTILE
Power	0.05	0.04
Labour	12.69	7.05
Gas	0.49	0.26
Consumable	0.64	2.75

Consumable/Process

■ Consumable   ▨ Gas
▢ Labour   ▨ Power

Costs of 300mm test weld calculated
using 'Symphony' spreadsheet.

Figure 4.12   Cost savings produced when a rutile flux-cored wire is used for vertical butt joints.

The level of both the particulate fume and, in particular, potentially toxic substances must be controlled to comply with health and safety requirements and this is normally achieved by means of simple local fume extraction.

(c) Consistency of the consumable

The manufacture of flux-cored consumables is more complex than the manufacture of solid wires but probably comparable with that involved in the production of MMA electrodes. It is important that the flux core is chemically homogeneous and evenly distributed throughout the consumable. It is also important that the surface of the wire is clean and free from excess drawing lubricant. Although these requirements presented a problem in early consumables the introduction of improved manufacturing techniques and on-line quality monitoring [50] now ensures that consistent consumable properties are maintained.

4.5 Summary

There have been steady improvements in consumables for MMA, SAW and GMAW welding. The introduction of high-toughness, moisture-resistant low-alloy steel and enhanced-performance rutile-coated stainless steel MMA electrodes, iron powder addition and Ti–Bo microalloyed wires for SAW and the use of low-residual wires for GMAW are useful examples of this evolutionary process. The introduction and exploitation of FCAW is perhaps more remarkable as indicated by the application trends which have been discussed in Chapter 2.

5 Gases for Advanced Welding Processes

5.1 Introduction

Shielding gases are an essential consumable in many of the more recently developed welding processes, for example:

(i) as the primary shielding medium in GTAW, GMAW, plasma and gas-shielded FCAW;

(ii) gases for laser formation, shielding and plasma control in laser welding.

5.2 Shielding Gases for Arc Welding Processes

The primary functions of the shielding gas in the arc welding processes are to provide a suitable medium for the stable operation of a sustained low-voltage arc and to provide shielding from atmospheric contamination. Secondary, but equally important, functions include the control of weld bead geometry and mechanical properties.

5.2.1 Arc support and stability

The arc is sustained by the flow of current in an ionized gas. The ease of ionization of the gas will therefore influence the ability to initiate and maintain the arc. The ease of ionization is indicated by the ionization potential of the gas and values for the common shielding gases are given in table 5.1. Argon in particular has a low ionization potential and is commonly used in the TIG process.

Table 5.1   Basic properties of common shielding gases.

ELEMENT	FIRST IONISATION POTENTIAL (ELECTRON VOLTS)	DENSITY (Kg/m³)
Argon	15.75	1.784
Helium	24.58	0.178
Hydrogen	13.59	0.083
Nitrogen	14.54	1.161
Oxygen	13.61	1.326
Carbon Dioxide	(14.0)	1.977

The thermal conductivity of the gas will also influence the arc stability; a high thermal conductivity may result in a reduction in the diameter of the conducting core of the arc and this can lead to an increase in voltage and a reduction in arc stability. Hydrogen, which has a lower ionization potential but a higher thermal conductivity than argon, increases the arc voltage when mixed with argon and may affect both arc stability and arc initiation if more than 8% is added.

Arc stiffness is usually regarded as beneficial in low-current GTAW arcs but as the current increases the force on the weld pool increases and this may result in undercut. It has been shown, however [51], that helium-rich gas mixtures produce a significantly lower arc force which may be beneficial at high currents (figure 5.1). Arc stability in GMAW and FCAW arcs is largely dependent on the mode of metal transfer which in turn is influenced by the effect of the gas on the surface tension and work function of the material and the resultant arc root behaviour.

5.2.2   Shielding from atmospheric contamination

The effectiveness of the gas in providing shielding from atmospheric contamination will depend on its chemical reactivity and its physical properties. In TIG welding it is necessary to protect the tungsten electrode from oxidation and for this reason the inert

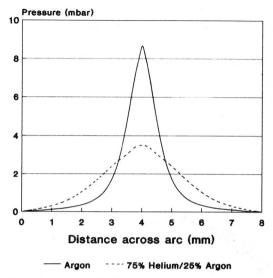

Figure 5.1 Pressure distribution for GTAW (150 A) with 3 mm arc length (after Al-Rawe A 1991 Welding and joining technology *MPhil* SIMS, Cranfield Institute of Technology).

gases argon or helium are normally used. In GMAW and FCAW welding it is usually necessary to use oxidizing additions to promote good metal transfer but the effect of these gases on the loss (by oxidation of metal droplets in the arc) of alloying elements must be considered.

It is also necessary to protect the weld metal from adverse gas–metal reactions such as the formation of porosity, inclusions, surface oxidation or embrittlement. Common active gases which may cause problems in this respect are oxygen, nitrogen and hydrogen. Most materials form oxides when heated in an oxidizing atmosphere whereas nitrogen may form insoluble nitrides with reactive metals (Ti, Ta, V, Nb) and soluble compounds with other metals (Fe, Mn, Cr, W). Hydrogen is soluble in most metals but can form compounds with the reactive metals.

The equilibrium solubility of both nitrogen and hydrogen is high in the liquid phase of many common metals but much lower in the solid (figure 5.2). If the molten metal absorbs more of these gases than the solid solubility limit there is a potential for porosity formation as the excess gas attempts to escape from the solidifying weld pool. Under arc welding conditions it is found that the amount of gas absorbed is higher than that which is expected under equilibrium conditions with non-arc melting [52] and, in the case of nitrogen, absorption in steel is affected by the alloying elements present and the level of oxygen and nitrogen. This is particularly interesting in terms of shielding efficiency since oxygen tends to increase the amount of nitrogen absorbed and hence, as recent work [53] has shown, the effect of air entrainment into the arc atmosphere is far more serious than the presence of nitrogen as a minor impurity in the gas.

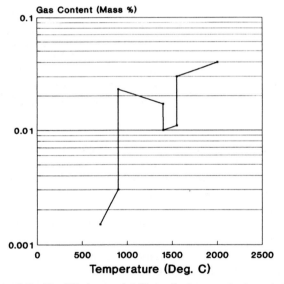

Figure 5.2 Equilibrium solubility of nitrogen in iron (after Lancaster J F 1987 *The Metallurgy of Welding* 4th edn (London: Allen and Unwin)).

The ability to maintain a lamellar gas flow and prevent atmospheric contamination will depend on the physical properties of the gas and in particular its density, viscosity and Reynolds number.

The compatibility of the common shielding gases with a range of materials is summarized in table 5.2.

Table 5.2 Compatibility of shielding gases with common materials.

SHIELDING GAS	COMPATIBLE WITH	PROBLEM AREA
Argon and Helium mixtures	All materials	None
Oxygen containing mixtures	Plain carbon and stainless Steels up to around 8%	Embrittlement of reactive metals (eg. Ti), oxidation poor weld profile and loss of alloying elements in most materials.
Carbon Dioxide	Plain carbon steel	Carbon pick up in ELC stainless steels.
Nitrogen	Copper	Porosity in steel and nickel, embrittlement of reactive metals, reduced toughness in alloy steels.
Hydrogen	Austenitic stainless steel and high nickel alloys up to around 5%	Porosity in aluminium and other materials. HICC in hardenable ferritic steels.

5.2.3   Secondary functions of the gas

The secondary characteristics of shielding gases are no less important than the primary functions and in some cases may determine the most suitable gas for a given application. Some of the most important secondary functions are control of fusion characteristics and joint properties.

(a)   Fusion characteristics

The shielding gas has a significant influence on the weld bead profile and fusion characteristics (figure 5.3). The total fused area is increased (at an equivalent current) by using gases which increase the arc energy (e.g. He, H_2, CO_2).

In GMAW welding the use of pure argon produces a pronounced 'finger' or 'wine-glass' penetration profile whereas argon/CO_2 and argon/helium mixtures produce a more rounded profile (figure 5.4).

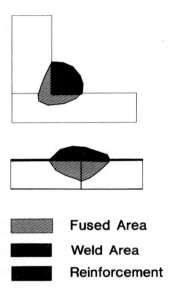

Fused Area

Weld Area

Reinforcement

Figure 5.3 Fusion geometry of fillet and square butt welds. The gas can affect the ratio of fusion area to total area, the fusion profile and the reinforcement geometry.

The profile of the reinforcement can also be improved; for example argon/CO_2 mixtures give flatter weld beads and consequent improvements in resistance to fatigue as well as cost savings (see also Chapter 7).

In some cases the shielding gas mixture may improve the consistency of fusion; argon/hydrogen mixtures have been found to have a beneficial effect on cast-to-cast variation in the GTA welding of austenitic stainless steel (as discussed in Chapter 6).

(b) *Joint properties*

The mechanical properties of the weld will depend on freedom from defects and the final weld metal microstructure, both of which are influenced by the shielding gas. Porosity may be controlled by selecting an appropriate shielding gas and ensuring that an efficient gas shield is maintained. Fusion defects may be minimized by selecting a gas which gives increased heat input and oxide inclusions may be limited by controlling the oxidizing potential of the gas.

(*a*)   (*b*)

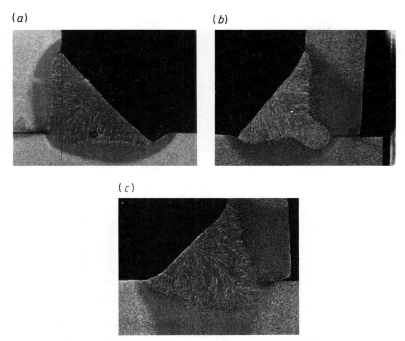

(*c*)

Figure 5.4   Bead profiles of fillet welds in 5 mm thick steel plate (GMAW).
(*a*) Poor bead profile, undercut and porosity produced when pure argon
is used. (*b*) Transfer and arc stabilized by the addition of 5% oxygen to
argon. Note the pronounced 'finger' penetration. (*c*) Stable transfer and
'bowl'-shaped penetration plus increased fusion when argon/20% CO_2 is
used.

The final weld microstructure may be influenced by the gas as a
result of its effect on heat input and weld metal composition. For
example it has been found [54] that with ferritic steels an improvement in toughness may be produced by increasing the oxidizing
potential of the GMAW shielding gas (by adding up to 2% oxygen
and 15% CO_2 to argon) as shown in figure 5.5. This is thought
to be due to the nucleation of fine-grained acicular ferrite by
controlled levels of microinclusions. A further increase in the oxidizing potential could, however, lead to the formation of coarse
oxide inclusions which would result in a deterioration of weld

metal toughness. These effects are relatively small and also rely on the composition of the welding wire and the ability to maintain stable process performance.

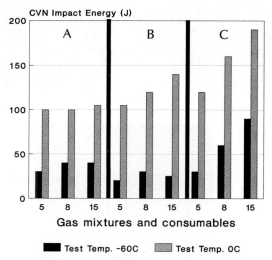

Figure 5.5 Toughness of high-strength, low-alloy steel welds made in various gas mixtures using pulsed GMAW. A, 70S6 wire 1; B, 70S6 wire 2; C, K5Ni wire: 5, 8, 15% CO_2 in argon. (After Norrish *et al* [54].)

5.2.4   Shielding gas options

The range of viable shielding gas options is limited by the need to satisfy the criteria listed above. Some of these gases are listed below.

(a)   *Argon*
Argon is one of the most widely used shielding gases for TIG welding. It is totally inert and has a high density relative to air. The low ionization potential facilitates arc striking and stability.

(b)   *Helium*
Helium is chemically inert, has a lower density than air and requires a higher arc voltage (at the same current and arc length)

than argon. The resultant increase in power produces increased heat input and fusion area although lower depth to width ratios are normally experienced. The cost of helium is considerably higher than that of argon but the welding speeds that are usually obtained make it a viable option, particularly for high-conductivity materials.

(c) *Carbon dioxide*

Carbon dioxide is chemically active but has a higher density than air. It can dissociate in the arc to release oxygen and carbon monoxide and this can result in a reduction in the weld metal content of elements such as silicon, manganese and titanium and an increase in carbon. Because of its chemical activity its use is restricted to GMAW welding of steel. The arc voltage is 1–2 V higher in CO_2 (for an equivalent current and arc length) than that found in argon-based mixtures and the heat input is slightly higher resulting in increased fusion. Transfer behaviour, operating tolerances and arc stability are generally poor, especially at high currents.

(d) *Oxygen*

Oxygen is not used as a shielding medium but is an important constituent of many gas mixtures. When added to argon it improves arc stability, reduces the surface tension of steel and improves arc root behaviour. The reduced surface tension improves metal transfer and bead shape. Like CO_2, the use of oxygen will decrease the recovery of the more reactive alloying elements.

(e) *Hydrogen*

Hydrogen increases the arc voltage and heat input when mixed with argon. Its use is usually restricted to the TIG process and to materials which do not suffer any adverse chemical or physical changes in its presence. Its chemically reducing properties may be used to advantage on austenitic stainless steels where it promotes wetting and produces improved weld bead finish.

5.2.5 Shielding gas mixtures for specific applications

By analysing the effects of the various individual gases it is possible to produce mixtures to satisfy the requirements of most material–process combinations. The composition and properties of the range of gas mixture available are described below.

(a)   Gas mixtures for GMAW welding of plain carbon and low-alloy steels

Carbon dioxide may be used for dip transfer GMAW, but mixtures based on argon with additions of oxygen and carbon dioxide are found to give improved arc stability, reduced spatter and an increased operating range (i.e. voltage, wire feed speed and inductance settings are less critical). In addition weld bead profile is improved, giving a saving in weld metal and weld time. The mixtures available normally fall into one of the following groups:

(i) argon plus 1–8% oxygen;
(ii) argon plus 1–8% carbon dioxide;
(iii) argon plus 8–15% carbon dioxide;
(iv) argon plus 15–25% carbon dioxide;
(v) pure carbon dioxide;
(vi) argon/carbon dioxide/oxygen mixtures.

Pure argon is unsuitable for GMAW welding since the arc is unstable and the resultant weld bead profile is irregular. The addition of less than 1% oxygen gives a remarkable improvement in arc stability, although the weld bead reinforcement is usually excessive and the penetration profile has a wine-glass appearance. Mixtures containing 1 to 2% oxygen may be used for pulsed spray transfer GMAW welding, but whilst arc stability is very good the penetration profile does not improve with the addition of oxygen. Higher oxygen levels, up to 8%, can be used for dip transfer welding of thin sheet. If more than 8% oxygen is added unacceptable surface oxidation may occur.

The addition of up to 8% CO_2 to argon promotes stable operation and gives a slightly improved bead shape, although the wine-glass penetration shape is still apparent. Mixtures containing about 5% CO_2 give smooth spray and pulsed transfer and low spatter levels in dip transfer. These mixtures are most suitable for welding thin material, for pulsed transfer welding of positional welds in thicker material or for high-current downhand and horizontal vertical fillet welds. The application of these low oxygen and carbon dioxide mixtures for multipass welding of thicker materials is limited by the sporadic appearance of fine inter-run porosity. This has been ascribed to argon entrapment and nitrogen absorption [55] and is known to be reduced by increasing the operating current or raising the CO_2 level of the gas.

Intermediate levels of CO_2 (argon plus 8–15% carbon dioxide) give a decreased risk of porosity, improved fusion and still maintain good operating characteristics in the spray and pulse transfer operating modes.

Mixtures of argon with 15–25% carbon dioxide are characterized by increased fusion with bowl-shape penetration profiles, but arc stability tends to decrease as the CO_2 level approaches 25%. These mixtures are ideally suited to the welding of thicker sections and for multirun butt welds. When the CO_2 content exceeds 25–35% the characteristics of the gas are similar to those of pure CO_2.

Pure CO_2 gives good fusion characteristics but higher weld bead reinforcement than the argon-rich mixtures. The heat input of the arc is increased and this may provide slightly better performance

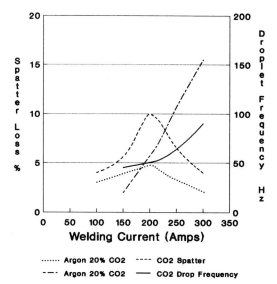

Figure 5.6 Comparison of spatter losses for CO_2 and argon/20% CO_2 mixtures (GMAW, 1.2 mm wire) (after Scheibner [56]).

on oxidized or primed plate. Although satisfactory dip transfer can be obtained in CO_2 the setting tolerances are narrower than those for argon-based mixtures and transfer under spray and pulsed conditions tends to be globular with much higher levels of spatter [56]. An indication of the difference in spatter levels associated with the use of CO_2 compared with an argon/CO_2 is shown in figure 5.6.

Ternary argon/carbon dioxide/oxygen mixtures have similar performance characteristics to argon/CO_2 but slightly improved arc stability has been observed. The results of arc stability measurements for dip transfer in these three-part mixtures is illustrated in figure 5.7, and it can be seen that mixtures containing 12–15% CO_2 and 2–3% O_2 give excellent stability. Mixtures in this range also give good spray transfer performance, even with electrode-negative operation [57] and good fusion characteristics.

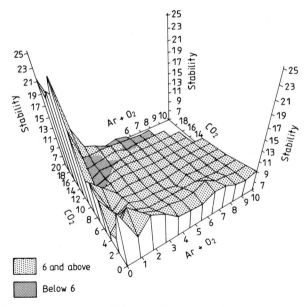

6 and above

Below 6

Figure 5.7 Arc stability, dip transfer GMAW in argon/CO_2/O_2 mixtures. (Stability is measured as the standard deviation of arc time, i.e. low values of the standard deviation indicate good stability.)

(b) Gas mixtures for GMAW welding of austenitic stainless steel

Stainless steel may be welded in the spray mode using argon with small additions of oxygen (1–2%), but if these mixtures are used for dip transfer operation bead appearance and fusion characteristics tend to be poor. Argon/5% CO_2 mixtures may be used to weld austenitic stainless steel but the carbon level of the resultant welds can increase above 0.04% making this mixture unsuitable for the low-carbon 'L' grade steels.

Helium additions to an argon/CO_2 mixture give improved fusion, reduced wetting angle and improved bead appearance, and several proprietary mixtures are based on mixtures of argon, helium and CO_2, with small additions of O_2 and hydrogen in some cases.

The common mixtures for stainless steel GMAW fall into two categories:

(i) high helium (60–80%);
(ii) low helium (20–40%).

The high-helium mixtures are used predominantly for dip transfer where the higher helium level increases welding speeds, improves bead appearance and increases the dip transfer frequency. The arc voltage is increased and fusion is improved especially at low currents.

Lower-helium mixtures have been developed mainly for spray and pulsed transfer welding. They promote smooth spray transfer, good fusion and excellent bead profiles. The addition of 1–2% hydrogen to these mixtures improves wetting and bead appearance by chemically reducing the surface oxide.

Although these mixtures were specifically developed for austenitic stainless steel they can be used for plain carbon steel where increased speed and improved surface appearance are required (e.g. in automated welding of thin sheet steel components).

(c) Gases for GTA welding of steels

Argon is the most widely used gas for GTAW although mixtures of argon with up to 5% hydrogen are often used, particularly for austenitic stainless steels where increased speed, improved profile and improved process tolerance are required. Hydrogen additions cannot be used on ferritic steels which are susceptible to hydrogen-induced cold cracking.

Helium/argon mixtures with 30–80% helium can be used for high-speed welding of steels, and on stainless steels both helium/argon and argon/hydrogen mixtures have been found to increase the tolerance to cast-to-cast variation problems [58].

(d) Gases for GMAW and GTA welding of aluminium alloys

Argon is normally recommended for both GMAW and GTA welding of aluminium and its alloys although mixtures of argon and helium with up to 80% helium offer improvements in fusion and bead profile [59,60]. These mixtures are particularly useful on thicker materials where the preparation angles and the number of weld runs can be reduced.

(e) Gases for GMAW and GTA welding of copper and its alloys

For materials such as copper with high thermal conductivity a higher heat input in the arc is desirable, particularly for GTAW. Helium or helium/argon mixtures give the necessary increase in heat input and reduce the need for preheat and/or give higher welding speeds and improved process tolerance. Nitrogen and nitrogen/argon mixtures have been used for GMAW; the nitrogen increases the heat input but transfer is poor and spatter levels can be high.

(f) Gases for GMAW and GTA welding of nickel and its alloys

Argon or argon/helium mixtures may be used for all nickel alloys. High-nickel alloys are susceptible to nitrogen porosity but small additions of hydrogen (1–5%) improve weld fluidity and reduce porosity. Argon/hydrogen mixtures are often used for GTA welding of the cupro nickels such as Monel (66% Ni, 31% CU).

(g) Gases for plasma welding

In plasma welding two gas supplies are required; the plasma gas and the shielding gas. For many applications the most suitable plasma gas is argon. It allows reliable arc initiation and protects the tungsten electrode and the anode orifice from erosion. The shielding gas may be argon, although for austenitic stainless steel additions of up to 8% hydrogen may be made to increase arc constriction, fusion characteristics and travel speed.

(h)   Gases for FCAW welding of steel

The gases used for the FCAW process are dependent on the type of consumable, i.e.:

(i) gases suitable for rutile and basic consumables, carbon and low-alloy steels;

(ii) gases suitable for metal-cored consumables, carbon and low-alloy steels;

(iii) gases suitable for stainless steel FCAW.

Most of the rutile and basic flux-cored wires are formulated to give good operating characteristics and mechanical properties with CO_2 shielding. The use of argon/20% CO_2 and the three-part argon/12–15% CO_2/2–3% O_2 mixtures usually gives a slight improvement in arc stability and improved recovery of easily oxidized alloying elements.

Metal-cored wires were originally designed to give smooth spray transfer in argon/5% CO_2 mixtures but the three-part argon/oxygen/carbon dioxide mixtures above give excellent transfer characteristics and slightly improved fusion. Metal-cored wires which operate satisfactorily with CO_2 shielding have recently been introduced.

Most of the stainless steel flux-cored wires available at present are designed to operate in argon/20% CO_2 mixtures but the three-part mixtures referred to above have also been found to be satisfactory. Recent work [61] has shown that the use of argon/CO_2 gas mixtures with lower levels of CO_2 improves the recovery of alloying elements, reduces weld metal oxygen levels and increases the yield and ultimate tensile strength of the weld. As with solid carbon steel wires optimum toughness was obtained with specific mixtures, and this is attributed to a combination of weld chemistry modification and the effect of the gas on the thermal cycle.

In the case of flux-cored wires the flux formulation will be designed on the assumption that a particular range of shielding gas mixtures will be used and it is important to ensure that any deviation from the intended gas mixture will not adversely affect the operating performance or weld properties.

(i) Gases for MIAB welding
MIAB welding can be performed without a shielding gas although it is reported that the use of CO_2 can offer some improvement in joint quality [62].

(j) Special mixtures
Certain special gas mixtures have been produced for specific applications. For example:

  (i) argon/chlorine, argon/Freon mixtures;
  (ii) argon/sulphur dioxide;
  (iii) argon/nitric oxide gas mixtures;
  (iv) gases for high-deposition GMAW.

Chlorine and Freon. Argon/chlorine mixtures have been investigated [63] as a means of reducing porosity and improving process tolerances in the GMA welding of aluminium. Although some improvements were reported the application of these mixtures is unlikely due to the extreme toxicity of chlorine.

Certain argon/Freon† mixtures are non-toxic and it has been found that the Freon may be substituted for chlorine to produce similar effects [64]; in particular arc stability and weld bead geometry were found to be improved. Although non-toxic the industrial exploitation of these mixtures is now restricted by environmental concerns.

In spite of the limitations on the use of both of these groups of mixtures it is possible that they may be useful for totally automated applications in totally enclosed controlled-environment chambers.

Sulphur dioxide. Argon/SO_2 mixtures have been used [65] to reduce the effect of cast-to-cast variation in GTA welding of austenitic stainless steel. Whilst the experiments offer a useful indi-

---

† Freons are gaseous combinations of carbon, chlorine, fluorine and bromine, e.g. Freon 12 is CCl_2F_2.

cation of the desirable influence of sulphur on the weldability of these materials, the toxicity of the gas precludes its use in practical applications.

Nitric oxide. A range of gas mixtures containing a small amount of nitric oxide (nitrogen monoxide, NO) has been developed as a means of controlling ozone levels in the vicinity of GMAW and GTAW arcs [66]. Ozone is formed by the irradiation of the oxygen in the air surrounding and in the immediate vicinity of the arc with ultraviolet light. Radiation with a wavelength in the 130–170 nm range is particularly effective in promoting ozone formation [67]. Ozone is toxic and the maximum recommended level in the welder's breathing zone is extremely low (0.01 PPM). Fortunately the ozone will react with other gases and substances in the atmosphere to form oxygen and oxides and at low currents or in the presence of a reasonable amount of particulate fume the rate of ozone formation is low and the likelihood of its recombination before reaching the welder's breathing zone is high. In certain applications with high-current GTAW arcs and particularly in the GMA welding of aluminium very high levels of ozone are, however, formed. If nitric oxide, which is much less toxic than ozone, is added to the shielding gas it will combine with the free ozone to form oxygen and nitrogen dioxide.

It has been shown that the use of NO additions of 0.03% to an argon/20% CO_2 mixture can significantly reduce ozone formation in the GMA welding of steel [68,69] as shown in figure 5.8 and are effective in controlling ozone in GMAW and GTAW welding of aluminium. These gas mixtures are usually recommended in combination with local ventilation to ensure the removal of all traces of toxic gases from the welding environment.

High-deposition GMAW. High deposition rates may be achieved with electrode negative operation with the aid of argon/O_2/CO_2 gas mixtures as discussed above. Alternatively, special gas mixtures which when used with extended electrical stick-out give very high deposition rates (up to 15 kg h^{-1}) have also been developed, these mixtures are based on argon/helium/CO_2 and are intended for automated applications. A special torch design is required to ensure adequate shielding with the increased electrical extension [70].

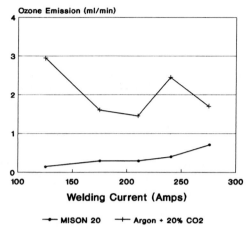

Figure 5.8 Effect of addition of nitric oxide on ozone formation during GMAW welding of steel. (Composition of MISON 20 is: argon/20% CO_2/max. 0.03% NO.) (From [68], courtesy AGA.)

5.3 Gases for laser welding

Gases are required in the laser welding process for operation of the laser, shielding and plasma control.

In gas or CO_2 lasers a gas mixture is used to support the electrical discharge and generate the laser beam. The exact mixture will depend on the type and manufacturer of the laser but typical gas mixtures are [71]:

(i) 80% helium/15% nitrogen/5% carbon dioxide;
(ii) 61% helium/4% carbon dioxide/31.5% nitrogen/
3.5% oxygen.

The way in which these gases are supplied (i.e. separate or premixed) will also depend on the type of gas laser being used. For both gas and solid state YAG lasers additional gases are required for shielding and plasma control.

For shielding the gases used are similar to those employed for GTAW and plasma welding but in the case of laser welding ioniza-

Table 5.3 Common   shielding   gases   for   arc   welding
processes.

GAS	APPLICATIONS	FEATURES
Argon	GTAW all metals, GMAW Spray/pulse Al, Ni, Cu.	GTAW; Good arc initiation stable arc. Efficient shielding, low cost. Poor bead profile in GMAW.
Helium	GTAW all metals; especially Cu, Al. GMAW high current spray.	High heat input, improved fusion, high arc voltage.
Argon + 25 to 80%Helium	GTAW and GMAW Al, & Cu	Improved fusion & bead profile.
Argon + 0.5 to 15% Hydrogen	GTAW austenitic stainless steel & Cu/Ni alloys.	Improved fusion and edge wetting, reduced oxides.
Carbon Dioxide	GMAW plain carbon and low alloy steels. Particularly dip transfer.	Low cost, good fusion, effective shield, but process stability poor and spatter high.
Argon + 1 to 7%CO_2 + up to 3%O_2	GMAW plain carbon and low alloy steel. Spray & pulse.	Low heat input, stable transfer, finger penetration.
Argon + 8 t0 15%CO_2 + up to 4%O_2	GMAW plain carbon and low alloy steel. Dip, spray and pulse.	Good arc stability, improved fusion and bead profile.
Argon + 16 to 25%CO_2	GMAW plain carbon and low alloy steel. Dip and FCAW.	Increased fusion but reduced process stability, & increased spatter.
Argon + 1 to 8%O_2	GMAW plain carbon, low alloy and stainless steel. Dip, pulse and spray.	Lower O_2 for pulse and spray. Reduced carbon pick up in stainless steels. Increased surface slag & oxide.
Helium + 10 t0 20% argon + O_2 + CO_2	GMAW dip transfer stainless steel.	Improved fusion and bead profile.
Helium + 30 to 40%argon + O_2 + CO_2 (or 0% CO_2 for low carbon pick up)	GMAW dip, spray and pulse stainless, plain carbon and low alloy steel.	Good fusion on all steels, can use at high current with high deposition GMAW.

tion of the gas or metal vapour to form a plasma is undesirable (see Chapter 8) and gases with a high ionization potential, such as helium, are favoured.

Common gas mixtures used for shielding are:

(i) argon, helium and argon/helium mixtures, used for most materials including steel and the reactive metals titanium and zirconium;

(ii) nitrogen can be used for less demanding applications on austenitic stainless steel.

If a plasma does form a jet of gas may be used to displace or disrupt the plasma [72], the normal gas used for this purpose is helium.

5.4 Summary

The range of gases used for shielding in arc and laser welding processes is limited but gas mixtures containing from two to four active components may be used to obtain the optimum welding performance. The range of gases commonly used for gas-shielded arc welding and their applications are summarized in table 5.3. The gas mixture selected will have a significant effect on the quality and economics of the resultant weld and the high basic cost of some of these mixtures must be evaluated against the overall cost of process as discussed in Chapter 2.

6 Advanced GTAW

6.1 Introduction

The GTAW process is well established as a high-quality fusion welding technique. Developments in the process have extended the potential application range and offer improved process control.

6.2 Process Developments

Some of the more significant process advances are described below and further developments in the field of automation, computer control and adaptive control are discussed in Chapter 11.

The basic process developments and the principles involved in the control of the process will be discussed under the following headings:

(i) arc initiation and electrode development;
(ii) pulsed GTAW and high-frequency pulsed GTAW;
(iii) cold- and hot-wire feed additions;
(iv) dual-gas GTAW and plasma welding;
(v) square wave AC GTAW and plasma;
(vi) multicathode GTAW;
(vii) control of GTAW and related processes.

6.2.1 Arc initiation and electrode development

Arc initiation in the GTAW process is a two-stage process consisting of initial breakdown of the arc gap (stage I) and stabilization of the arc (stage II). Stage I is influenced by the electrode, the open-circuit voltage of the power supply and the striking technique used. Stage II is controlled mainly by the rate of response of the power supply.

Initial breakdown of the arc gap may be achieved by one of the following:

  (i) touch striking;
  (ii) high-voltage DC;
  (iii) high-frequency high-voltage.

Each of these techniques has limitations, as discussed below and this has led to the development of new systems which are also described.

(a) *Touch striking*

This is probably the simplest technique available. The electrode is brought into contact with the workpiece then rapidly withdrawn. The tungsten electrode is heated resistively by the short-circuit current of the power source and the initial arc is established from the heated electrode immediately the electrode is withdrawn. The striking process is assisted by the very small arc length which exists the instant the electrode contact with the workpiece is broken and the presence in this gap of metal vapour. The effectiveness of the technique depends to a large extent on the skill of the operator but there is always a likelihood of tungsten contamination occurring and this will adversely affect the electrode running performance and weld quality.

(b) *High-voltage DC*

It has been shown that to obtain reliable arc breakdown at normal arc lengths DC voltages of 10 kV would be required [73]. These voltages would pose serious safety hazards and are not feasible for normal applications. The use of short-duration high-voltage surges has been shown to reduce the danger of lethal electric shock [74] but there is still a risk of injury from reaction to accidental contact with such high pulse voltages. The application of this technique is therefore restricted to automatic systems in which the operator is protected from contact with the high-voltage supply.

(c) High-frequency–high-voltage

High-frequency currents tend to be carried in the outer layers of a conductor and this 'skin' effect can be used to advantage in GTAW arc striking systems. High-frequency–high-voltage supplies (e.g. 3 kV at 5 MHz) are effective in breaking down the arc gap and are non-injurious to the operator. This type of system has been used extensively for arc starting and AC arc stabilization in GTAW and continues to be the main method of striking used in manual GTAW systems. In some cases it is found that arc starting using high-frequency ignition systems becomes inconsistent, this may be due to electrode characteristics or an adverse phase relationship (i.e. lack of synchronization) between the power supply and the arc starting device. It has also been suggested that a negative space charge may be generated around the end of the electrode and the gas cup. In this case improved consistency may be obtained by discharging the charge by connecting the gas cup to the positive terminal of the power supply. Where a ceramic gas cup is used a conductive foil may be wrapped around the nozzle to provide an electrical connection, whereas with metallic cups a resistor is inserted between the cup and the positive connection.

The main problem with the high-frequency starting technique, however, lies in the use of high-frequency voltage oscillations which, depending on the design of the oscillator circuit, can cover a wide range of radio frequencies and produce both airborne and mains-borne interference. In the past this problem has resulted in interference with communication systems and domestic television and radio reception but it is also likely to create significant problems with electronic control and computing equipment in the welding environment.

(d) New arc striking techniques

Programmed touch striking. The main problem with conventional touch striking is the high short-circuit current which tends to overheat the electrode and increase the risk of contamination. This limitation can be overcome by controlling the current during the short circuit. Various systems exist [75] but the operation is essentially as follows (see also figure 6.1). After closing the torch switch (A) a low voltage is applied between the electrode and the workpiece via a current-limiting resistor, when the electrode touches the workpiece (B) the short circuit is detected electronically and a low current (2–10 A) is allowed to flow; this current is sufficient to preheat the electrode without overheating (C). When the electrode is lifted the voltage rises (D) and signals the power supply to initiate the main current supply. The initial arc current may be programmed to rise rapidly to ensure arc stabilization before dropping to the working value. Trials have shown [76] that no evidence of tungsten contamination or electrode weight loss could be detected after repeated re-striking with a system of this type. The system is also ideally suited to automatic application where the contact of the electrode with the workpiece and its retraction can be mechanized (see Chapter 11).

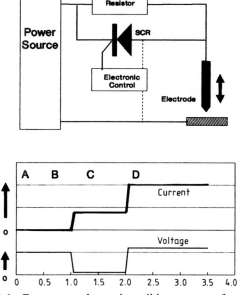

Figure 6.1 Programmed touch striking system for GTAW. Top: simplified circuit diagram. Bottom: current and voltage during programmed touch striking.

Pilot arc starting. The use of an auxiliary electrode in the torch enables a low-current pilot arc to be struck before initiation of the main arc. This system allows consistent striking although it does require a slightly more complex torch.

Piezoelectric arc starting. Piezoelectric arc starting devices have been investigated [77] and it has been shown that a torch-mounted piezoelectric device can be used successfully for GTAW arc starting. Problems were, however, encountered with high-voltage leakage and the system has not yet been exploited commercially.

(e)   *Arc stabilization*

Stage II arc initiation or stabilization is largely dependent on the rate at which the power source can supply current to the embryo arc after initial breakdown. Conventional GTAW power sources have been shown to have a current response of around $10^4\,\mathrm{A\,s^{-1}}$ at the optimum settings and with a favourable phase relationship [78], although much lower rates of rise could be experienced, particularly on single-phase units when the arc initiation takes place out of phase with the mains supply. It would be expected that this rate of rise of current could lead to striking difficulties especially at low set currents and when long, inductive welding cables are used. In these circumstances effective arc initiation often depends on the use of a capacitor in parallel with the output which can discharge into the arc.

Electronically controlled power sources, such as the series regulator and inverter designs described in Chapter 3 are capable of giving much higher rates of initial current rise (typically $5 \times 10^4\,\mathrm{A\,s^{-1}}$) and are less sensitive to phase relationship problems.

(f)   *Electrode composition*

The tungsen electrodes used in GTAW are usually alloyed with a small amount of thoria or zirconia in order to improve arc starting, by reducing the work function of the tungsten and improving its emission characteristics.†

Thoriated electrodes give very good striking and DC running characteristics but it has been demonstrated that the consistency of performance is closely related to the homogeneity of the electrode,

---

† The work function of pure tungsten is around 4.54 eV whereas that of a 2% thoriated electrode is around 2.63 eV.

and in particular the regularity of the thoria distribution [79]. In this study it was shown that the stable arc operating time (continuous arc operation) may be extended by up to 100% when an electrode with a fine, homogeneous distribution of thoria particles (70 h stable arc operation at 125 A) is substituted for an electrode with a less regular composition (35 h stable operation), whilst re-ignition delays can be reduced from 4% of the total number of arc starts to 1% with the better-quality electrode.

Although thoria (ThO_2) is effective in improving arc striking and tip shape retention it is naturally radioactive. Concern about potential safety implications, in particular in electrode manufacture, has led to the investigation of alternative alloying additions. Oxides of the rare earth elements lanthanum, yttrium and cerium appear to offer similar characteristics to thoria. Laboratory investigations [80] indicate that electrodes doped with these substances may perform better than conventional thoriated types. In these tests the number of successful arc initiations using high-frequency (HF) arc starting and an open-circuit voltage (OCV) range of 18 to 36 V was assessed. The results are summarized below and in figure 6.2.

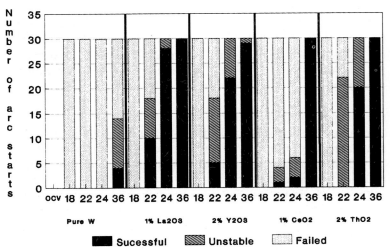

Figure 6.2 Electrode performance (arc start rate versus OCV) for various GTAW electrodes with rare earth additions. (After [80].)

Follow on current was set to 20–30 A. The electrode vertex angle was 45°, and the total number of attempts at each open circuit voltage was 30. At 30 V OCV the performance of ThO_2, La_2O_3, CeO_2, and Y_2O_3, is very similar, whereas at 24 V OCV the lanthanum oxide gave the best results. Measurements of electrode temperature indicated that La, Y and Ce oxides gave lower operating temperatures than those for pure tungsten and zirconiated tungsten. Similarly the amount of electrode melting and tip shape deterioration was much less when La, Y, and Ce oxide additions were used.

(g) *Rim formation and weight loss*

A rim of tungsten 'whiskers' forms on the vertex of the electrode, particularly if there is more than 0.05% oxygen present in the shielding gas. This effect is thought to be associated with the volatilization of tungsten oxide and the condensation and growth of pure tungsten crystals on the cooler part of the electrode. This rim can lead to arc asymmetry and instability. The effect is not directly affected by the alloying addition, although weight loss of the electrode was less significant with the rare-earth-doped electrodes.

The improvements in electrode life and striking performance obtained with rare-earth-doped electrodes are relatively small but where reliable and consistent operation is required, for example in automated welding, these marginal improvements may prove beneficial. Alternatively high-quality thoriated electrodes may be specified in situations where consistency is important.

6.3   Process Variants

6.3.1   Pulsed GTAW

Low-frequency (1–10 Hz) modulation of the current in the GTAW process has been used [81] to provide the following process characteristics:

(i) reduced distortion;
(ii) improved tolerance to dissimilar thicknesses;
(iii) improved tolerance to dissimilar materials;

(iv) reduced heat build up;
(v) improved tolerance to cast-to-cast variations.

Using electronic power sources it is possible to generate a range of alternative pulse profiles although in practice nominally square wave pulses are usually used as illustrated diagrammatically in figure 6.3. The low current or background level is set at a value just sufficient to maintain an arc without causing significant plate melting. The ideal pulse current level is determined by the thermal properties of the material [82] and should be set at a level which will ensure that the weld pool can propagate at a rate which is fast enough to ensure the maximum thermal efficiency.

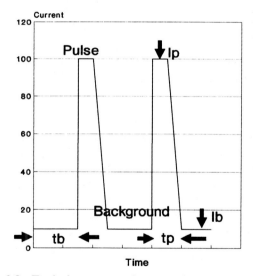

Figure 6.3 Typical GTAW pulse waveform: variation of current with time.

Table 6.1 gives some guidance on the optimum values for a range of materials. The duration of the pulse is determined by the thickness of the material and if the pulse current exceeds the values given in table 6.1 a relationship of the form

$$I_p t_p = K \qquad (6.1)$$

should hold, where I_p is pulse current, t_p is pulse time and K is a constant. The background duration is adjusted to allow solidification to occur between pulses and this will in turn be related to the travel speed.

The pulsed GTAW process has been applied widely for demanding applications such as the joining of austenitic stainless steel for cryogenic expansion bellows, orbital welding of nuclear reprocessing plant pipework and the welding of aeroengine components in nickel-based super-alloys (Nimonics).

Table 6.1   Operating parameters for pulsed GTAW.

Material	Pulse amplitude (amps)
Pure nickel	250 to 300amps
Stainless Steel	150 to 200amps
Cuppro-nickel alloys	150 to 200amps
Plain carbon steel	100 to 150amps
Nimonic alloys	50 to 80amps

6.3.2   High-frequency pulsed GTAW

High-frequency (above 5000 Hz) pulsed GTAW has been investigated as a method of improving arc stiffness, giving high energy density and efficiency [83] and enabling higher welding speeds to be achieved. The improved arc stiffness is claimed to be most significant at low average arc currents and at frequencies in the range 5 kHz to above 20 kHz. The arc stiffness is likely to be due to the high pulse current amplitude, the high-frequency pulsing between this high peak and very low background level allowing a low mean current to be maintained. At higher mean currents high-frequency pulsing gives an increased resistance to undercut and this enables the maximum welding speed to be increased in mechanized welding applications.

6.3.3   Square wave AC GTAW

AC GTAW welding is the recognized operating mode for aluminium and its alloys. However, self-rectification effects in the arc make it necessary to provide DC component suppression and some means of assisting re-ignition of the arc at every current zero. Limitations on allowable open-circuit voltages and radio interference from the high-voltage–high-frequency units make it difficult and costly to design conventional GTAW equipment to overcome these problems.

The developments in power source technology already discussed in Chapter 3 have made it possible to produce a 'square' output waveform instead of the normal sine wave [84–86]. This has several beneficial effects on the process. Firstly the rapid polarity reversal at current zero reduces the degree of cooling and recombination in the arc and assists the arc to re-strike more easily in the opposite direction. This re-ignition process may be further assisted by arranging for high-voltage transients to coincide with current zero. Improved control of the process is also provided by the ability to adjust the balance between alternate half cycles. It is, for example, possible to increase the arc cleaning effect by increasing the relative duration of positive (electrode) half cycles whilst plate fusion may be increased by extending the negative half cycle. These effects are shown in figure 6.4. In addition to the benefits described

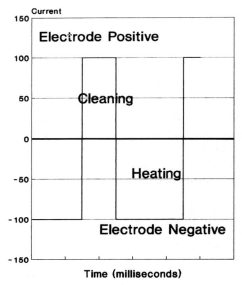

Figure 6.4 The effect of square wave AC operation on GTAW.

above it is normal for this type of unit to have mains voltage stabilization plus low-voltage remote control.

A further development of this technology is the use of high-frequency transistor reversing switch circuits which, when added to the output of a conventional or inverter-based DC power source, enable a square wave AC output to be obtained. It is reported [87] that a unit of this type is capable of reversing the polarity of the output within 100 μs and gives effective cathodic cleaning on aluminium alloys with 20 ms electrode negative polarity and 4 ms positive (i.e. a pulse frequency of 41.66 Hz.)

6.3.4 Cold- and hot-wire additions in GTAW

A useful feature of the GTAW process is the ability to control weld pool temperature and size by means of independent cold-wire addition. This is used in manual welding operations to control the weld profile and penetration, particularly in positional work. Mechanized wire addition by means of a separate hand-held wire dispenser or a torch-mounted feed system are available for manual operation, but although these can increase the operating efficiency of the process they tend to be restricted to simple weld geometry and joints with good access.

Figure 6.5 'Dabber' TIG welding system for rebuilding the cutting edge of an end mill: left, workpiece; centre, torch; right, filler wire delivery system.

In automated GTAW the wire is normally fed at a set constant rate and the element of control exerted by the welder is lost. A mechanized wire-feed addition system called 'Dabber GTAW'†, which feeds the wire with a reciprocating motion similar to that used by manual welders has been introduced [88]. The frequency of wire addition may be set between 2 and 10 Hz. A controlled low heat input deposit may be achieved and the technique is particularly useful for building up fine edges of cutting tools (figure 6.5) or components such as turbine blades.

6.3.5 Hot-wire GTAW

Normally GTAW is regarded as a 'low-productivity–high-quality' process due to the relatively slow travel speeds employed and low deposition rate achieved with cold filler additions. It has been shown, however, that significant improvements in deposition rate, to match those produced in the GMAW process, may be achieved by using a 'hot-wire' addition [89,90].

The process arrangement is illustrated in figure 6.6. The principal features are the addition of a continuously fed filler wire which is resistively heated by AC or DC current passing between the contact tip and the weld pool. Normally the wire is fed into the rear of the weld pool (unlike cold-wire addition) although hot-wire

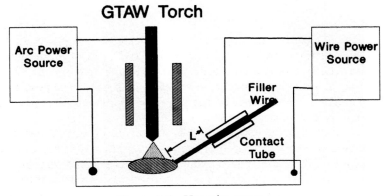

Figure 6.6 Hot-wire GTAW.

† Dabber TIG is a trade name of the Hobart Company Ltd.

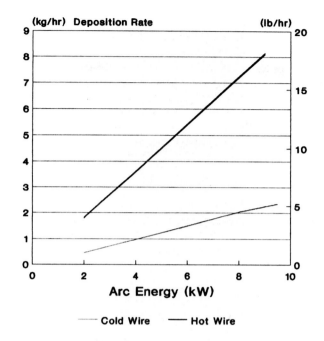

After Manz.A.F. WRC Bulletin No.223.1977

Figure 6.7 Comparison of deposition rates for hot- and cold-wire GTAW. (After Manz [89].)

additions to the front of the pool have been used for positional work. The equipment required comprises a precision power source (e.g. electronic control mains-voltage stabilized), a high-quality wire-feed system and an effective gas-shielding system. The use of an AC power source for filler wire heating minimizes the possibility of magnetic disturbance of the arc. The deposition rates possible are shown in figure 6.7.

A novel system based on an inverter power source has been developed for evaluation in the power generation industry; this consists of a DC inverter-based GTAW power source which supplies the arc power. An additional power circuit and control system is powered from the high-frequency AC output of the inverter and provides the heating supply to the wire. This design is cost effective and electrically efficient and compact as shown in figure 6.8.

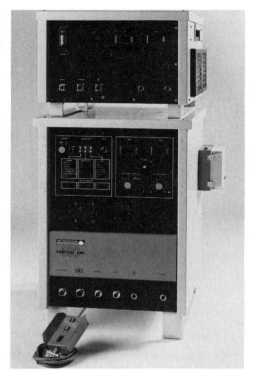

Figure 6.8 Hybrid inverter/phase-controlled regulator for hot-wire GTAW. (From Wright [91].)

Although the equipment is more complex than GMAW it has been shown that deposition rates of $10-14 \, \text{kg h}^{-1}$ are possible and high joint integrity may be expected. The hot-wire GTAW process has been applied in the oil industry for butt welding 30 mm wall line pipe. In this application four heads rotate around the pipe simultaneously as shown in figure 6.9.

6.3.6 Dual-gas GTAW

Constriction of the core of a GTAW arc occurs under normal operating conditions due to the effect of the compressive Lorentz force which is produced by the interaction of the arc current and

Joint detail

Welding parameters		
Pass	Current (A)	Wire feed speed (m min^{-1})
Root	220	3
Hot	350	5
Fill	360	7
Cap	250	5.5

1.2 mm dia. wire
75% helium/25% argon

Figure 6.9 Multihead hot-wire GTAW system for line pipe welding. (After Belloni A and Carossio G 1985 A new generation, fully automatic welding system *Proc. 1st Int. Conf. 'Advanced Welding Systems', London 1985* (London: The Welding Institute) paper 14, pp. 1–11. (See also figure 9.6.)

its associated magnetic field. The Lorentz force is proportional to the square of the arc current and the resultant constriction may be very low at currents below 20 A but pronounced at currents above 100 A. The effect can be induced by applying an external axial magnetic field or by thermal constriction caused by the impingement of cold gas jets on the outer region of the arc. Reducing the temperature of the outer core of the arc decreases the area available for current flow, restricts the number of charge carriers (ionized particles) and the temperature and energy density of the inner core of the arc must increase.

This effect has been utilized in the 'dual-shielding' GTAW technique [92] illustrated in figure 6.10. A cylindrical nozzle surrounding the electrode directs a flow of cool shielding gas along the outer surface of the arc. This gas provides shielding of the electrode and the immediate arc area but also causes some thermal constriction and stiffening of the arc. An additional concentric gas shield provides protection of the weld pool and outer regions of the arc. The gases used for the inner and outer shields may be of different com-

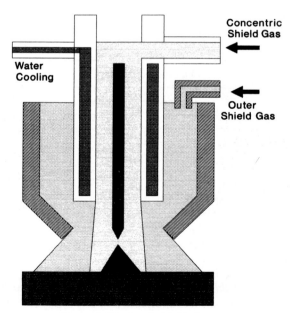

Figure 6.10   Dual-shielded GTAW.

positions, for example argon/5% hydrogen may be used for the central gas (giving a potential increase in constriction and stiffness (see Chapter 4) while argon or even argon/20% CO_2 may be used for the outer shield when welding low-carbon steel. The dual-gas system has been used for stainless steel, carbon steel and non-ferrous materials including aluminium. At low currents (20 to 50 A) the process gives improved arc stability and the mean current required is 30 to 40% lower than that required for conventional GTAW. On thicknesses up to 3 mm in aluminium alloys and 4 mm in steel, square butt preparations may be used and welding speed increases of up to 20% compared with conventional GTAW are possible. If the current is increased to 335 A and above keyhole† welding is possible on plate thicknesses from 4 to 6 mm.

Physical constriction of the GTAW arc by a water-cooled copper nozzle as shown in figure 6.11 will produce an even higher current density and arc core temperature for a given mean current. When this technique is used the process is referred to as the plasma process [93] (see also Chapter 1).

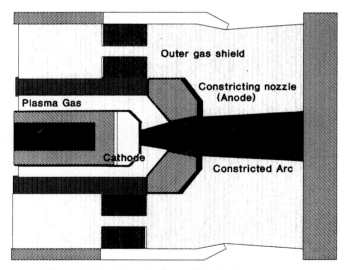

Figure 6.11   Constriction of arc in plasma torch.

† See Chapter 8 for further information on the keyhole mode of operation.

6.3.7   Plasma welding

In plasma welding the arc is formed between the tip of a non-consumable electrode and either the workpiece or the constricting nozzle. Again it is possible to select different shielding and plasma-forming gases, although argon is commonly used as the central plasma gas whilst argon or argon/hydrogen (where appropriate) may be used for the outer shielding medium.

Two basic operating modes are possible for plasma welding. These are:

(i) the transferred arc mode;
(ii) the non-transferred arc mode.

(a)   *The transferred arc mode*

In the transferred arc mode the arc is maintained between the electrode and the workpiece as shown in figure 6.12. The electrode is usually the cathode and the workpiece is connected to the positive side of the power supply. In this mode a high energy density is achieved and this energy is efficiently transferred to the workpiece.

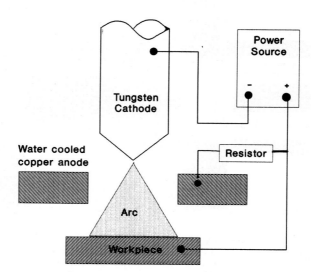

Figure 6.12   Transferred arc mode.

(b) *The non-transferred arc mode*

In this mode of operation the power supply is applied to the electrode and the constricting orifice. The electrode is usually connected to the negative side of the power supply and for this reason the
orifice is referred to as the anode. This process mode is illustrated
schematically in figure 6.13. In this mode very little energy is
transferred to the workpiece and its main use is as a 'pilot' arc
which enables the main transferred arc to be established rapidly.
The pilot arc may be established by high-frequency arc starting
or a simple touch-starting system within the torch. Once the pilot
arc is established a main or transferred arc may be started at any
time by completing the circuit between the power source and the
workpiece as shown in the diagrams above.

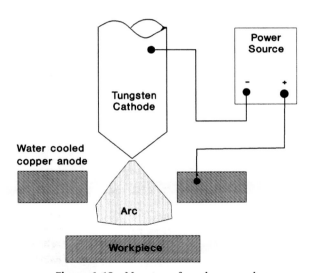

Figure 6.13   Non-transferred arc mode.

(c) *Equipment requirements*

The basic equipment requirements for the plasma welding system
are slightly more complex than either the conventional or dual-gas
GTAW system. They consist of a welding power source, a control
unit and a plasma torch.

A power source with a direct current (DC) output is normally
used (although it is possible with suitable equipment to use AC for

aluminium alloys and square wave AC units referred to above have been developed as plasma power supplies). The constant-current static volt−amp output characteristic used for GTAW power sources provides a suitable main arc supply.

The control unit which usually contains the non-transferred pilot arc supply, the gas controls and the arc ignition system may take the form of an 'add on' box or may be built into the main power supply.

The torch is slightly more complex than a GTAW torch as shown in figure 6.14. It must have provision for cooling of the replaceable copper constricting orifice, supply of plasma gas, and a separate supply of shielding gas. The non-consumable electrode material is usually thoriated tungsten. As in GTAW the process may be used without added filler (autogenous welding) but if a filler is required this is usually added to the leading edge of the molten pool.

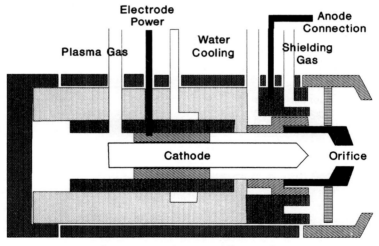

Figure 6.14 Plasma welding torch.

(d) *Modes of operation of plasma welding*

The plasma welding process is normally operated with direct current electrode negative (DCEN). Although, as with GTAW, both AC and DCEP operation can be used for aluminium and its alloys, this usually requires special torch designs and larger electrodes. Welding may be carried out in the 'melt in' mode in a similar manner to GTAW, the only difference is the use of a constricted arc as the heat source. Alternatively the keyhole mode may be used; this takes advantage of the higher energy density and the increased arc force in the plasma and is described in more detail in Chapter 8.

(e) *Features and applications of plasma welding*
The features and applications of plasma welding differ according to the current range used, i.e.:

(i) low current: 0.1–15 A;
(ii) intermediate current: 15–200 A;
(iii) high current: above 200 A.

Low-current or 'microplasma' welding. In the low-current range the principal advantage of the process over GTAW is the excellent arc stability even at levels of 1–2 A at which it would be difficult to operate a GTAW arc. Restricted arc root area and improved directionality are added advantages. At low currents the pilot arc ensures reliable main arc starting and can also be used to illuminate the joint prior to welding. These features of the process make it suitable for joining very thin materials, for the encapsulation of electronic components and sensors, the joining of fine-mesh filter elements [94] or repair of turbine blades as shown in figure 6.15.

Intermediate-current plasma welding. In the range 15–200 A the process is similar in characteristics to GTAW but for the same mean current higher speed or improved melting efficiency may be achieved. In addition the plasma process is more tolerant to variations in stand-off, and the tungsten electrode is protected from accidental touch-down and contamination. The process has been used in this range for the joining of silicon iron transformer laminations where contamination of the electrode presents problems in the GTAW process.

High-current plasma operation. At high currents the keyhole mode of operation may be used. This makes it possible to perform

Figure 6.15 Microplasma welding used for repairing an aircraft turbine engine blade; the original part was manufactured with EB technology. (Courtesy Messrs Griesheim GmbH.)

single-pass square butt welds from one side in plate thicknesses up to about 9 mm. Accurate alignment of the torch and stable travel speed are important in this mode to avoid undercut. Typical applications of keyhole plasma welding are high-speed longitudinal welds on strip and pipe. The process has also been used for root runs in thick-wall pipe.

6.3.8   Multicathode GTAW

One of the principal factors which limits the maximum welding speed of the GTAW process is the occurrence of undercut [95] (see section 6.4 below) and although dual-shield and plasma techniques increase the potential heating efficiency of the arc they also have a tendency to increase the risk of undercut, and this in turn limits their usefulness for high-speed welding. This problem may be overcome by creating an elongated heat source using two or more arcs in series (figure 6.16). This technique is called multicathode GTAW and was originally developed more than 20 years ago [96]. The welding speed obtained is determined by the number of electrodes

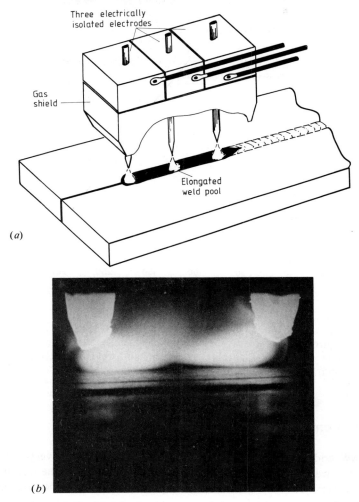

Figure 6.16 Multicathode GTAW: (*a*) general arrangement; (*b*) arc appearance—two electrodes 12 mm apart.

and their separation but, as shown in figure 6.17, these are significantly higher than the travel speeds obtained with conventional GTAW. This process variant has been used mainly for specialized applications in the high-speed welding of the longitudinal seams in tube mills. In order to avoid the tendency for the arcs to deflect under the influence of the local magnetic fields (figure 6.16) techniques such as high-frequency pulsing, dual-shield and magnetic stabilization are used.

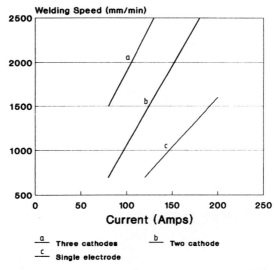

Figure 6.17 Increases in welding speed achieved using multiple electrodes for full-penetration butt welds. (Material used: 1.2 mm 304-type stainless steel.)

6.4 Control of GTAW and Related Processes

GTAW and related processes are capable of producing very high-quality welds but for consistent results the influence of the welding parameters on weld geometry and quality must be identified and controlled.

6.4.1 Conventional GTAW

In conventional DC GTAW the main control parameters are:

Primary	Secondary
Current	Arc length
Travel speed	Polarity
	Shielding gas
	Electrode vertex angle
	Filler addition

(a) *Current and travel speed*

The mean current normally determines the heating effect of the arc and the arc pressure and stiffness. The penetration and fusion characteristics for a fixed set of secondary parameters are determined by a combination of mean current and travel speed as shown in figure 6.18. However, the maximum speed, penetration and fusion area are limited by the onset of unacceptable weld bead profile at high currents. The force generated in the high-current arc displaces the molten weld pool and at high travel speeds there is insufficient time for the displaced metal to flow into the joint prior to solidification. The result is a discontinuous bead, undercut or 'humping' as shown in figure 6.19. The occurrence of this defect

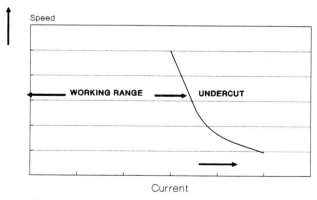

Figure 6.18 Speed/current relationship for GTAW.

Figure 6.19 Weld bead humping in GTAW—longitudinal section through bead on plate weld.

has been investigated by various workers [97] and the limiting current and speed have been determined for a range of conditions as shown in figure 6.20. The effect of this limitation may be reduced by controlling some of the secondary variables or using multicathode techniques.

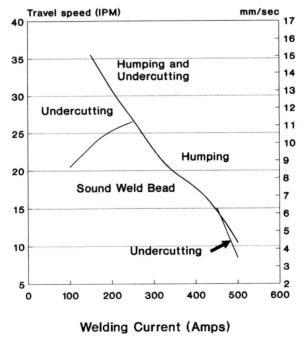

Figure 6.20 Incidence of humping and undercut. (After Savage *et al* [97].)

(b) Arc length

The arc length in GTAW is usually taken to be the same as the separation distance between the electrode tip and the workpiece. The effect of increasing this distance is to decrease the heating efficiency of the arc and the fusion and penetration level. This reduced efficiency which is due to radiation losses from the arc column occurs even though arc voltage increases (see figure 6.21) and the total power (current × voltage) may increase. For consistent results with conventional GTAW it is therefore important to maintain a fixed electrode-to-workpiece separation. The arc voltage gives a useful indication of electrode-to-workpiece distance.

(c) Polarity

The thermal balance between the anode and cathode in GTAW is such that some 60–70% of the heat input to the electrodes is absorbed by the anode whilst only 30–40% is absorbed at the cathode [98]. For most applications electrode negative polarity is used since this provides the best heating characteristics and minimum electrode/torch heating. For aluminium and its alloys electrode positive polarity offers one important benefit in providing cathodic cleaning of the plate surface, but in order to optimize the efficiency of the process alternating current is normally used for these materials; this provides plate heating during the electrode negative half cycle and plate cleaning during the electrode positive period.

(d) Shielding gas

The shielding gas can have a significant effect on the thermal characteristics of the arc and fusion behaviour of GTAW as discussed in Chapter 5. These effects can be used to extend the operating range of the process and, for example, the maximum speed before the onset of weld bead humping at 400 A may be increased from 7.6 mm s^{-1} to 23.3 mm s^{-1} by replacing argon with helium as the shielding medium.

(e) Electrode vertex angle

The angle at which the tungsten electrode is ground has a marked effect on the arc pressure (figure 6.22). Small angles increase the arc pressure and arc voltage and large included angles reduce the pressure. For the avoidance of undercut and humping at high

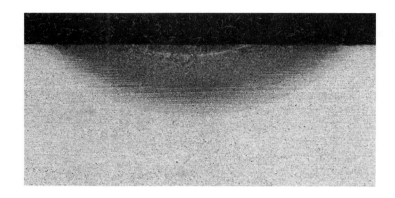

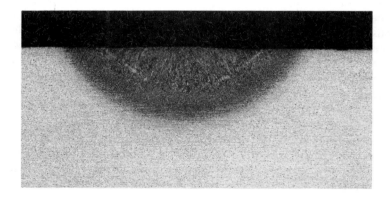

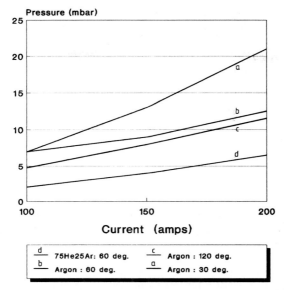

Figure 6.22 Effect of vertex angle on arc pressure. (After Al-Rawe A 1991 Welding and joining technology *MPhil* SIMS, Cranfield Institute of Technology.)

currents it is therefore preferable to use a large vertex angle. The vertex angle also has an influence on weld bead geometry at much lower currents, the effect is related to relative† plate thickness and is summarized in table 6.2. For a high depth to width ratio on thin plate (or deep penetration welds) a small vertex angle is desirable whereas for high depth to width ratios on thick plate, or low

† Relative plate thickness is the thickness of the plate relative to the penetration achieved: 'thick' plate means a penetration of less than 30%, 'medium' where 30–70% penetration is achieved and 'thin' plate represents the 70–100% penetration situation.

Figure 6.21 (Opposite) The influence of arc length on plate fusion in GTAW. All tests were done at 150 A, travel speed 5 mm s^{-1} and DC electrode negative. Top: arc length 8 mm, voltage 13.8 V, arc power 2070 W; middle: arc length 5 mm, voltage 11.25 V, arc power 1688 W; bottom: arc length 2 mm, voltage 9 V, arc power 1350 W.

penetration welds, a larger angle is required. These effects are, however, related to material composition, shielding gas and joint geometry, and whilst it is clear that there is a significant change in penetration characteristics with any change in electrode angle the resultant weld profile will depend on the material and the other welding variables discussed above.

Table 6.2   Penetration profile in relation to electrode vertex angle.

Electrode vertex angle (degrees)	Bead profile (depth to width ratio) as a function of relative plate thickness		
	Thin Plate	Medium	Thick plate
30	High	Low	Low
120	Low	High	High

(f) *Filler addition*
The addition of cold filler wire will cool the weld pool and reduce the heat available for plate fusion; however in some circumstances small traces of elements which alter the surface tension of the weld pool may be added via the filler to improve the fusion characteristics as discussed below.

(g) *Cast-to-cast variation in GTAW*
Although control of the GTAW process is straightforward if the variables listed above are considered, variable weldability has been experienced, particularly in the fabrication of stainless steel at relatively low currents. This phenomenon is referred to as cast-to-cast variation since it commonly occurs when a new batch of material of nominally identical composition is used. The problem has received considerable attention and is believed to be associated with the level of trace elements in the material and, in the case of austenitic stainless steel, sulphur, calcium and oxygen variations have been identified as having a major influence. For example the two welds shown in figure 6.23 were made under identical welding conditions on low-carbon austenitic stainless steel (304L), the only difference in the analysis of the plate material was in the level of sulphur which was 0.004% in the case of the low depth to width ratio weld (*a*) and 0.007% in the case of weld (*b*).

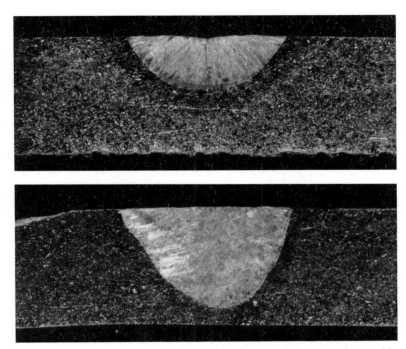

Figure 6.23 Cast-to-cast variation: sections of two GTAW spot welds made under identical conditions on different casts of 304L stainless steel.

The influence of such small fluctuations of elements such as sulphur has been explained in terms of their influence on the surface tension–temperature gradient of the liquid weld pool and the subsequent flow within the pool. It is known that the surface tension gradient of molten iron can be altered by the presence of trace elements as shown in figure 6.24 [101]. If the slope of the surface tension–temperature curve is negative (the surface tension is higher at low temperatures) surface-tension-driven flow of liquid metal will take place across the surface of the pool from the high-temperature region at the centre to the low-temperature area at the outer edge as shown in figure 6.25(a). If the surface tension gradient has a positive slope the metal will flow inwards from the periphery of the pool, and the higher-temperature liquid will flow down the axis to promote increased central fusion (figure 6.25(b)).

This effect is known as 'Marangoni flow' [102]. The mechanism would appear to explain the sensitivity of many materials to variations in trace elements and the resultant weld geometry.

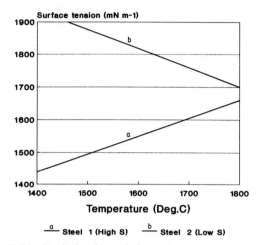

Figure 6.24   Variation in surface tension–temperature curves for iron and the influence of trace elements. (After Mills K C and Keene B J 1990 *Int. Mater. Rev.* **35**(4).)

The penetration may also be influenced by other factors: electromagnetic or Lorentz forces, buoyancy (flow induced by density differences in the pool due to the temperature gradient), and aerodynamic drag forces (due to the flow of gas/plasma jets over the pool surface) are all thought to have an effect. At higher currents the energy transfer and flow within the pool are likely to be dominated by the effect of the arc plasma and depression of the pool surface.

The effects of cast-to-cast variability can be reduced by using higher currents, pulsed operation, by choice of shielding gas (e.g. argon/5% hydrogen for austenitic stainless steel) or by adding a filler containing elements which promote a positive surface tension–temperature gradient (e.g. sulphur in austenitic stainless steel). It has also been reported that the problem may be alleviated by coating the material with a surface-active paste.

In order to control the problem, however, it is necessary to identify potentially problematic materials. Several approaches may be adopted, i.e. chemical analysis, direct weldability trials or indirect weldability trials.

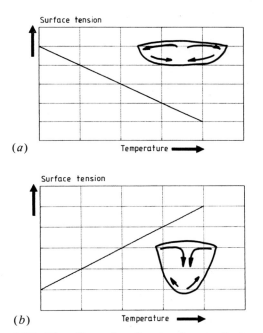

Figure 6.25 The effect of surface tension gradients on weld pool flow.

Chemical analysis may be a useful indicator of potential problems but it may be costly and unreliable in view of the low levels of surface-active elements which need to be identified and the uncertainty of the correlation between the analysis and the penetration profile.

Direct weldability trials involve making a joint of the required type using an established welding procedure, sectioning the bead and measuring its cross sectional area. This approach has been common but it is time consuming and requires expert analysis.

A variety of indirect weldability tests have been investigated [103–105] and it has been found that it is feasible to identify the cast-to-cast effect by measuring the time for a weld to penetrate a known thickness of plate under controlled conditions; alternatively the material characteristics may be identified from direct observation of the weld pool using video techniques or by monitoring light and voltage signals from the arc.

6.4.2 Control of pulsed TIG

In pulsed GTAW the process is controlled by the variables described above for conventional GTAW plus the pulse parameters. The influence of the pulse parameters has been described above.

6.4.3 Control of plasma welding

The control of plasma welding is slightly more complex than GTAW. The main control variables are the same as GTAW but plasma gas flow rate and the diameter and geometry of the plasma orifice will also have a significant effect on the operation of the process. A smaller plasma orifice will produce an increased arc force whereas a large nozzle diameter will result in a 'soft' plasma which is more like a GTAW arc. If the plasma gas flow is too low, the current too high, or the nozzle cooling is restricted an arc may form directly across the gap between the electrode and the nozzle, this 'double arcing' phenomenon will result in serious damage to the orifice.

As with GTAW, higher currents allow higher travel speeds to be used but as the current is increased there is a more noticeable

Table 6.3 Principal control parameters for GTAW and plasma processes.

Control parameters GTAW
Current, pulse parameters, current rise/decay times, electrode polarity, welding speed, electrode geometry, shielding gas type, shielding nozzle size, shielding gas flow stabilisation (gas lens), shielding gas flow rate, electrode protrusion.
Additional control parameters for PLASMA
Electrode set back, nozzle geometry, orifice diameter, plasma gas type, plasma gas flow, pilot arc current.

increase in arc force which may result in undercut. If both the plasma gas flow and current are increased the keyhole mode of operation is possible as described in Chapter 8.

The main factors controlling GTAW and related processes are summarized in table 6.3.

6.5 Summary

The capabilities of GTAW have been extended by basic modifications to the operating technique such as pulsing, multicathode and hot-wire addition, the use of new shielding gases and power sources, and fundamental variations such as the dual-gas and plasma welding systems. The application of these techniques can offer improved weld quality and increased productivity. A better understanding of the way in which the process control variables influence joint quality has been developed and this has enabled further advances to be made in the automation of the process. Some of the resultant developments will be discussed in Chapter 11.

7 Gas Metal Arc Welding

7.1 Introduction

Due to its high operating factor and deposition rate GMAW has the potential to improve productivity over that obtained with the GTAW and SMAW processes. Although the requirement to exploit the economic benefits of the process has led to a clear trend towards greater use of GMAW worldwide it has in the past proved difficult to obtain reliable quality. The main thrust of development has therefore been to improve control and achieve more consistent quality.

In order to discuss the advances which have been made in this direction it is necessary to reconsider the process fundamentals, in particular metal transfer in GMAW and control of conventional GMAW.

7.2 Metal Transfer in GMAW

The way in which material is transferred from the tip of the consumable electrode into the weld pool has a significant influence on the overall performance of GMAW: it affects process stability, spatter generation, weld quality and the positional capabilities of the process. Phenomenological studies of the mode of metal transfer have been carried out using high-speed cine or stroboscopic cine and video techniques [106]. The various types of transfer have been classified into groups. A simplified basic classification is shown in figure 7.1.

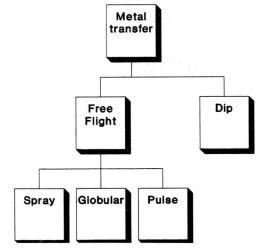

Figure 7.1   Classification of metal transfer in GMAW.

In free-flight transfer an arc is maintained between the electrode and the workpiece and the metal is transferred across the arc in the form of droplets. The size of the droplets and frequency of transfer may vary considerably and several subdivisions of free-flight transfer are necessary to accommodate these variations. The common free-flight modes are;

(i) globular (drop and repelled);
(ii) spray (drop projected streaming).

7.2.1   Globular drop transfer

Globular drop transfer is characterized by large droplets and low transfer rates (figure 7.2). It is normally found at low currents and fairly high arc voltages, although this will depend on the diameter of the filler wire, its composition and the shielding gas used. For example, with 1.6 mm diameter aluminium wire in an argon shield droplet transfer frequencies of less than 1 Hz may be observed at 100 A. In CO_2-shielded GMAW of steel, globular transfer occurs with a range of wire sizes at currents above 200 A.

Both the appearance of the droplet and observation of droplet formation indicate that the transfer mechanism is dominated by gravitational forces; i.e. the droplet detaches when its size has grown to a stage where the downward detachment force due to its mass overcomes the surface tension force which acts to prevent droplet separation. Although electromagnetic forces exist they are not sufficiently developed to influence the droplet detachment at low currents.

A low mean current is used but the process has very limited positional capabilities with solid wire GMAW because of the dominant nature of gravitational forces.

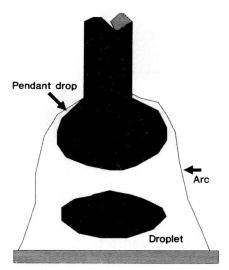

Figure 7.2   Globular transfer.

7.2.2   Globular repelled transfer

In some circumstances a droplet may form at the end of the electrode and be deflected to one side or even expelled from the arc. This behaviour is commonly found when electrode negative polarity is used with a solid wire and is illustrated in figure 7.3. The dominant transfer force is gravitational but repulsion is caused by electromagnetically induced plasma forces or vapour jets which act on the base of the droplet, at the arc root, to lift the molten material. Once the droplet has been lifted in this way an asymmetrical magnetic field is created and the droplet may be rotated or expelled under the influence of the resultant forces as discussed below. This mode of transfer is usually undesirable due to the poor stability and high spatter levels which result.

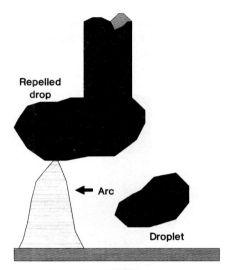

Figure 7.3   Repelled transfer.

7.2.3   Projected spray transfer

As the current is increased the size of the droplet usually decreases and the frequency of transfer increases. In addition it is found that the droplets are projected axially through the arc with some force. In some cases (e.g. carbon steel in argon-rich gas mixtures and aluminium in argon) there is a clear transition between the globular and projected spray modes of transfer as the current is increased (figure 7.4). The current at which this transition occurs is an important process characteristic and is known as the *spray transition current*. Its value depends on the filler material size and composition as well as the composition of the shielding gas. Typical values for steel are shown in table 7.1. Below the transition current the transfer is either globular or dip, and above the transition current the transfer is in the form of a steady stream of small droplets whose diameter is similar to that of the filler wire. Since this mode of transfer only occurs at relatively high currents the heat input is high and the weld pool large. These features are attractive for high-deposition-rate downhand welding but limit the positional capabilities of the process.

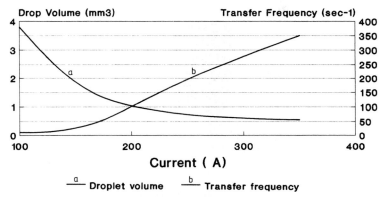

Figure 7.4 Transfer characteristics for 1.2 mm diameter stainless steel wire in argon/2% oxygen. (After Balraj V and Norrish J 1987 Pulsed MIG welding of stainless steel with controlled metal transfer *Proc. Ann. Conf. Indian Institute of Welding* (Madras: Indian Institute of Welding) pp. 893–907.)

Table 7.1 Spray transition currents for plain carbon steel wires.

Wire diameter in mm	Spray transition currents (amps) in various shielding gas mixtures		
	Argon/5%CO_2	Argon/15%CO_2	Argon/20%CO_2
0.8	140	155	160
1.0	180	200	200
1.2	240	260	275
1.6	280	280	280

7.2.4 Streaming transfer

As the current increases, the droplet size decreases further and the electrode tip becomes tapered. A very fine stream of droplets is projected axially through the arc as shown in figure 7.5. This mode of transfer is called *streaming* and it is caused by a significant increase in electromagnetic forces. It occurs more readily with high-resistivity, small-diameter wires (e.g. austenitic stainless steel) operating at currents above 300 A. Weld pool turbulence and gas entrainment may limit the usefulness of this mode of transfer.

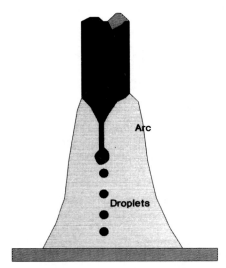

Figure 7.5   Streaming transfer.

7.2.5   Drop spray transfer

The transition to projected spray transfer occurs over a relatively narrow current range but it has been found [107] that an important intermediate transfer mode can occur in this transition range. This mode of transfer is characterized by the formation of a solid conic neck on the wire tip, and spherical droplets slightly larger in diameter than the diameter of the filler wire are initially suspended from the tip before being detached (figure 7.6). Detachment occurs very efficiently and high droplet velocities and very low spatter losses are measured. With a 1.2 mm carbon steel wire this transfer mode occurs between 250 and 270 A in argon/5% CO_2, drop velocities of 7 m min^{-1} have been recorded and a slight increase in melting rate is observed. The drop spray mode is efficient and 'clean' with very low spatter and particulate fume levels, but under normal steady DC operating conditions it requires very close control of the welding parameters and this can only be achieved with the high-quality electronic power sources described in Chapter 3; in addition the operating range is very restricted. The process range can, however, be extended by utilizing the pulsed transfer techniques described below.

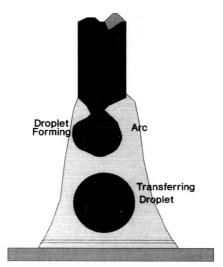

Figure 7.6 Drop spray transfer.

7.2.6 Dip transfer

If the electrode is fed toward the workpiece at a speed which exceeds the rate at which the arc alone can melt the wire, it will ultimately bridge the arc gap and dip into the pool. This behaviour may occur occasionally during free-flight transfer and is regarded as a fault condition but if the parameters are carefully chosen it is possible to induce regular short circuiting of the arc gap at frequencies above 100 Hz. This mode of transfer is known as *dip* or *short-arc transfer* and is illustrated diagrammatically in figure 7.7.

The sequence of operation is as follows. The wire is fed at a constant speed but burn-off during the arcing period is insufficient to

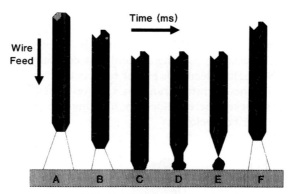

Figure 7.7 Mechanism of dip transfer.

maintain a constant arc length. The arc gap closes and the wire eventually contacts the weld pool. In response to this electrical short circuit, the current from the power supply rises rapidly causing resistive heating in the thin filament of wire which bridges the gap. The bridge ruptures, a portion of the heated electrode material is transferred to the weld pool and the arc is re-established. If the wire feed rate and power source output are carefully matched the short-circuiting process is repeated at regular intervals. The typical form of transient voltage and current waveforms is shown in figure 7.8. In practice the arc must remain relatively short (2–3 mm) to maintain a regular, high dip frequency and it is not uncommon for upward movements in the weld pool to initiate the short circuit.

High currents (typically 200–400 A) are required to rupture the short circuit but the arcing current is low and the arcing time is usually longer than the short-circuit time, as a result the mean current is maintained at a low level. If the current during the short circuit is excessive the short circuit will rupture explosively and metal will be ejected from the arc as spatter. During normal operation of the process there is some uncertainty concerning the exact amount of metal detached during each short circuit. The time between short circuits, the arc time, and hence the frequency of transfer, vary but the interval between short circuits usually follows a normal distribution as shown in figure 7.9. The standard deviation of the distribution may be used as an indicator of

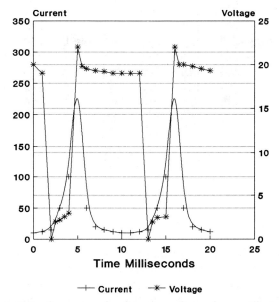

Figure 7.8 Current and voltage waveforms in one dip transfer mode.

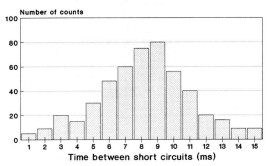

Figure 7.9 Histogram of time between short circuits (arc time) for stable dip transfer.

process stability, optimum conditions having the lowest value of standard deviation in arc time.

The low mean current, low heat input and resultant small, fast-freezing weld pool make dip transfer suitable for positional work and the welding of thin sheet steel.

The random nature of the short circuiting together with process instability and risk of high spatter levels are potential limitations of dip transfer. These problems can, however, be controlled either by choice of shielding gases (see Chapter 5) or by electronic techniques as discussed below.

7.2.7 Other transfer phenomena

Metal transfer may usually be classified into one of the categories above but several variations of normal transfer do occur.

(a) *Explosive droplet transfer*

It has been observed that pendant drops on the electrode tip can eject material in an explosive manner. This is thought to be due to chemical (gas–metal or slag–metal) reactions inside the droplet. These explosions may assist transfer in FCAW but usually cause instability in GMAW.

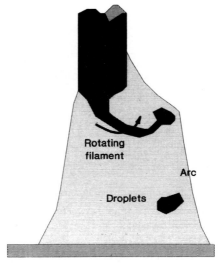

Figure 7.10 Rotating transfer.

(b) *Rotating transfer*

Rotation of the droplet may occur in the repelled mode as described above but the term *rotating transfer* is also used to describe the rotation of an extended metal filament between the solid wire tip and the droplet in streaming transfer as shown in figure 7.10. The occurrence of this mode of transfer at high currents is usually undesirable although it has been used for surfacing applications using the plasma–MIG process.

7.2.8 FCAW and slag protected transfer

In flux-cored wires the slag formed from the flux constituents may affect the transfer phenomena. The type of transfer depends on the flux system used but the following transfer phenomena have been identified [108,109] using a high-speed image converter camera. These effects are illustrated in figure 7.11.

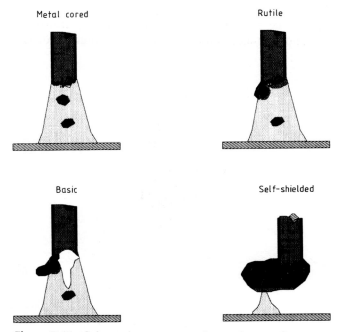

Figure 7.11 Schematic representation of transfer types observed with flux-cored wires.

(a) *Metal-cored wires*

These consumables contain very little non-metallic flux and tend to behave like solid wires. Good dip transfer performance is obtained at low currents and axial projected spray at higher currents. In addition stable electrode negative operation may be achieved in argon-rich argon/CO_2 gas mixtures. The streaming spray transfer which occurs at high currents (e.g. 350 A for a 1.2 mm diameter wire) gives high burn-off rates and smooth weld bead profiles.

(b) *Rutile flux-cored wires*

These consumables are normally operated in the spray mode where they give smooth non axial transfer. Some of the flux melts to form a slag layer on the surface of the droplet, a small amount decomposes to form shielding gases whilst some unmelted flux is transferred to the weld pool where it melts and produces a protective slag blanket. The unmelted flux projects from the tip of the wire as shown in figure 7.11.

(c) *Basic flux-cored wires*

The basic flux formulation gives irregular dip transfer at low currents and non-axial globular transfer at higher currents. The

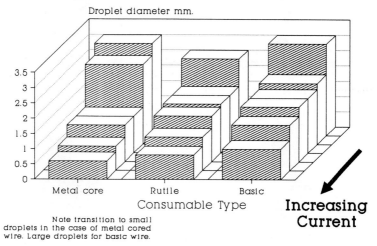

Figure 7.12 Droplet diameter range using image analysis.

unmelted flux forms a pronounced finger which projects from the wire into the arc. The effect of the flux formulation on the droplet size for gas-shielded flux-cored wires is shown in figure 7.12.

(d) Self-shielded flux-cored wires

Dip and globular repelled transfer are commonly found with this type of consumable and very large levitated globular 'boots' may form at the wire tip. The globular tendency may be reduced by flux formulation and there is evidence of secondary transfer occurring from the base of the globule as well as explosive droplet transfer.

7.3 The Physics of Metal Transfer

In order to understand and improve transfer behaviour it is necessary to consider the mechanisms involved in more detail. The transfer behaviour described above is a result of a balance of forces acting on the metal droplet. The principal forces involved are:

(i) gravitational force, F_g
(ii) aerodynamic drag, F_d
(iii) electromagnetic, F_{em}
(iv) surface tension, F_{st}
(v) vapour jet forces, F_v.

The dominant forces and their influence on metal transfer will depend on the operating conditions (current, voltage, wire diameter, shielding gas, etc) used but in free-flight transfer the balance of forces at the point of droplet detachment is illustrated diagrammatically in figure 7.13 and described by an equation of the form

$$F_g + F_d + F_{em} = F_{st} + F_v. \qquad (7.1)$$

In dip transfer the surface tension force may act to assist detachment and with the exception of F_{em} the other forces may be quite small.

An indication of the magnitude of these forces is given below, based on the commonly accepted classical physics approach, but many of the parameters involved are time and temperature dependent and a full theoretical analysis requires a consideration of the dynamic phenomena.

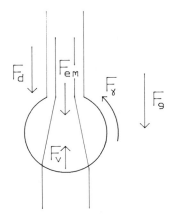

Figure 7.13   Balance of forces acting on a droplet.

7.3.1 Gravitational force

The gravitational force is given by

$$F_g = mg \qquad (7.2)$$

where m is the mass of the droplet and g is the vertical component of the acceleration due to gravity (i.e. $9.81 \cos \theta \ \mathrm{m \, s^{-1}}$ where θ is the angle between the arc axis and vertical). The force will have a maximum positive value in downhand welding (when $\cos \theta$ is $+1$) and a negative value in positional welding (when θ is between $90°$ and $180°$ and $\cos \theta$ has a negative value). Measured values of this force [110] for 1.6 mm wires in argon shielding gas at low currents (globular transfer) indicate values of 260 dyn for aluminium and 600 dyn for iron.

7.3.2 Aerodynamic drag

The gas flow within the arc can induce a force on a droplet, F_d, which may be calculated from;

$$F_d = 0.5 \pi V^2 dr^2 C \qquad (7.3)$$

where V is the gas velocity, d is the gas density, r is the droplet radius and C is the drag coefficient. The magnitude of this force will be highest when the droplet diameter and gas velocity are high. It is unusual for both gas velocity and droplet diameter to be at their maximum values at the same time; large droplets are normally found at low currents whilst high gas velocities are more commonly experienced at higher currents; drag forces are consequently small in most situations.

7.3.3 Electromagnetic forces

When a current flows through a conductor a magnetic field is produced and electromagnetic forces will be generated. The magnitude of these forces in the area surrounding the tip of the electrode, the molten droplet and the arc is strongly influenced by the geometry of the current path. The magnitude of the force may be calculated from

$$F_{em} = \frac{\mu I^2}{4\pi} \ln \left| \frac{r_a{}^2}{R} \right| \tag{7.4}$$

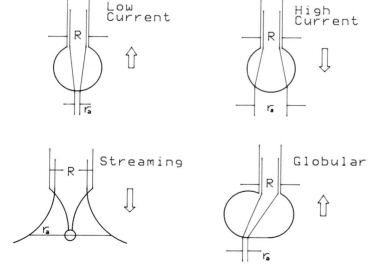

Figure 7.14 'Geometry' of current path in various transfer modes.

where μ is the magnetic permittivity of the material, I is the current, r_a is the 'exit' radius of the current and R is the 'entry' radius (see figure 7.14). Electromagnetic forces can have relatively large values (e.g. up to $0.02 \times I^2$ have been measured in GTAW arcs) and are clearly dependent on the current. These forces tend to dominate the transfer at the currents required for spray transfer.

7.3.4 Vapour jet forces

At high currents significant vaporization of the surface of the molten droplet can occur in the arc root area. Thermal acceleration of the vapour particles into the arc plasma results in a force which opposes droplet transfer. The value of this force for a flat surface of uniform temperature and composition can be shown to be

$$F_v = \frac{m_0}{d_v} \, IJ \tag{7.5}$$

where m_0 is the total mass vaporized per second per amp, I is the current, J is the current density and d_v is the vapour density. The vaporization force usually only becomes significant at higher currents or when low-vapour-pressure elements are present.

7.3.5 Surface tension

Surface tension plays a very important role in metal transfer; in free-flight transfer it is the principal force which prevents droplet detachment, and in dip transfer it is the major force which pulls the droplet into the weld pool. A simple static analysis of the drop-retaining force in globular transfer would suggest that the force is given by

$$F_{st} = 2\pi r_w \sigma f(r_a/c) \tag{7.6}$$

where r_w is the wire diameter, δ is the surface tension and $f(r_w/c)$ is a function of wire diameter and the constant of capillarity c. For large droplets the value of this equation approximates to $2\pi r_w \sigma$.

Calculation of the magnitude of the force due to surface tension is, however, complicated by the significant temperature dependence and the dramatic influence of certain surface-active elements (for example at the melting point of steel its surface tension will be reduced by around 30% by a concentration of 0.1% oxygen and the effect of small amounts of sulphur in changing the surface tension/temperature gradient has already been discussed in Chapter 6). Values of 300 dyn for aluminium and 600 dyn for steel have been calculated, however, for globular transfer with a 1.6 mm diameter wire.

7.4   Summary: Metal Transfer Phenomena

Metal transfer phenomena may be classified as free-flight or dip and within the free-flight mode several alternative transfer types may be observed. A classification which embraces these phenomena has been devised by the International Institute of Welding [111] and this is illustrated in table 7.2. The mode of metal transfer is influenced by a balance of forces which will depend on the operating parameters for the process. Gravitational, electromagnetic and surface tension are the most significant forces controlling metal transfer. In conventional GMAW the level of these forces and the resultant transfer behaviour is determined by the physical properties of the system (material and the shielding gas) but is controlled to a significant extent by the welding current.

7.5   Control of Conventional GMAW

Mean current determines the transfer mechanism of the process as described above and also controls the melting rate of the filler wire.

7.5.1   Melting rate phenomena: GMAW

The melting rate, MR, is usually expressed as

$$\mathrm{MR} = \alpha I + \frac{\beta l I^2}{a} \qquad (7.7)$$

where I is the current, l is the electrical stick-out or extension (figure 7.15) and a is the cross sectional area of the wire. α and β are constants.†

† Measured values of α and β for 1.2 mm plain carbon steel wire are $\alpha \approx 0.3$ mm A^{-1}s^{-1} and $\beta \approx 5 \times 10^{-5}$A^{-2}s^{-1};   for   aluminium $\alpha \approx 0.75$ mm A^{-1}s^{-1} and β is negligible.

Table 7.2   Classification of transfer modes (modified version of IIW classification) [111].

Transfer Group	Sub-Group	Example
1.0 Free Flight		
1.1 Globular	1.1.1 Globular drop	Low current GMAW
	1.1.2 Globular repelled	CO_2 shielded GMAW
1.2 Spray	1.2.1 Projected	GMAW above spray transition
	1.2.2 Streaming	Medium to high current GMAW
	1.2.3 Rotating	High current, extended stick-out GMAW
	1.2.3 Explosive	SMAW
	1.2.4* Drop spray	Pulsed GMAW
2.0 Bridging transfer		
2.1 Short circuiting		Low current GMAW
2.2 Bridging without interruption		Welding with filler wire addition
3.0 Slag protected transfer		
3.1 Flux wall guided		SAW
3.2 Other modes		SMAW,FCAW

  * Note: authors modification to IIW classification.

The first term in equation (7.7) represents the arc heating effect whilst the second term is due to resistive heating of the electrode. Melting rates are affected significantly by the electrical resistance of the stick-out as shown in figure 7.16. The stick-out resistance depends on the electrode diameter/cross sectional area, electrode resistivity, and the length of the extension. DCEN (direct current electrode negative) operation increases the melting rate [112] as shown in figure 7.17 but it is normally difficult to maintain a stable arc and ensure adequate fusion with this mode of operation.

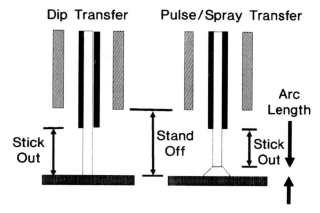

Figure 7.15   Terms used to describe torch position.

If the electrode polarity and the extension are fixed then for stable operation of the process in any transfer mode the wire must be fed at a rate (the burn-off rate) which is equal to the rate at which it is consumed (i.e. the melting rate †). The relationship

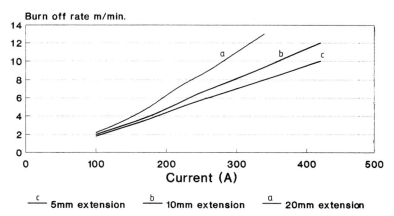

Figure 7.16   Influence of electrode extension (stick-out) on the burn-off rate for 1.2 mm diameter steel in argon/5% CO_2. (After Nunes J 1982 *MSc Thesis* Cranfield Institute of Technology.)

† The term *melting rate* is usually used to describe the mass of electrode material consumed per unit time. *Burn-off rate* is the rate at which the wire is consumed or the wire feed speed. *Melting speed* is sometimes used to describe the speed at which the melting isotherm or solid–liquid boundary travels along the electrode wire.

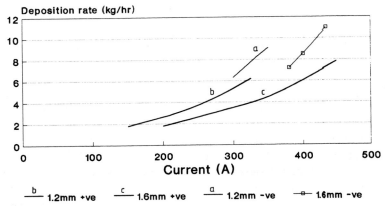

Figure 7.17   Influence of electrode polarity on deposition rate for steel in Ar/O$_2$/CO$_2$ mixture. (After Norrish [112].)

between wire feed speed and current which is given by equation (7.7) is usually shown graphically in the form of burn-off curves of the type shown in figure 7.18 and this allows the appropriate wire feed speed to be selected for a given mean current.

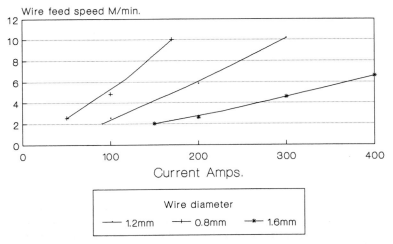

Figure 7.18   Burn-off characteristics for plain carbon steel filler wire.

7.5.2   Voltage–current characteristics

The voltage developed between the end of the contact tip and the workpiece in the GMAW process is the sum of the resistive drop in the wire extension plus the voltage fall across the arc. Calculation of the resistance of the electrode stick-out is complicated by the temperature dependence of resistivity and the steep temperature gradient which exists in the wire. Measurements of the total voltage drop under a range of operating conditions show that the relationship between mean current and voltage in the free-flight operating modes of the GMAW process is very similar to the characteristic of a GTAW arc. In the working range the arc has a positive resistance and for any shielding gas–filler wire combination at a fixed arc length the voltage increases linearly with current.

In dip transfer the mean current–voltage characteristic represents the average of the short-circuit resistance and the arc resistance and follows the same trend. In both dip and free-flight transfer the relationship between mean current and voltage may be expressed in an equation of the form

$$V = M + AI \qquad (7.8)$$

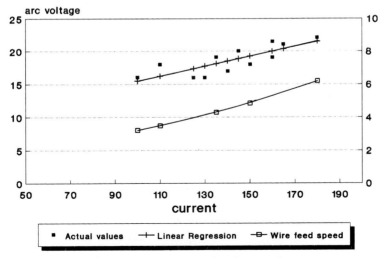

Figure 7.19   1.0 mm wire dip transfer.

where V is the arc voltage, I is the current and M and A are constants. The resultant relationship is illustrated in figure 7.19.

7.5.3   Control in conventional GMAW systems

Conventional wire feed systems are designed to maintain the feed speed constant at a preset value irrespective of any variations in the arc behaviour. Conventional GMAW power sources have for some time been designed with constant-voltage (CV) characteristics in order to provide self-adjustment and stabilization of the arc length. With these systems if the arc length tends to change, the current varies significantly and the burn-off behaviour acts in such a way as to counteract the change in arc length. An increase in arc length causes an increase in arc voltage and the power source output current must reduce in order to meet the higher voltage demand; since melting rate is current dependent the reduced current will result in reduced melting and since less wire is consumed the arc length is shortened. Shortening the arc will produce an increase in current, increased melting and again the arc length will return to its original value (figure 7.20).

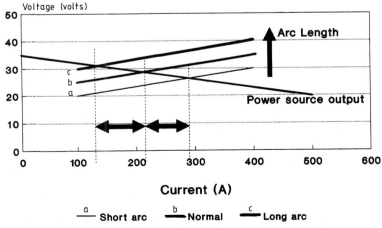

Figure 7.20   Self-adjustment with a constant-voltage power source.

7.5.4 Welding parameter selection: CV power source

Current is normally chosen according to plate thickness being welded and the travel speed required. The wire feed speed may be determined from the burn-off curves as described above and the voltage is set to provide the required current and satisfy the arc voltage–current characteristic. The operating point for the process will be defined by the intersection of the power source and the process characteristics, as shown in figure 7.21. The practical effects of incorrect adjustment are reflected in arc performance. If the voltage is too low for a given wire feed setting 'stubbing' † occurs. If the voltage is too high a long arc will occur with eventual burn-back of the filler wire to the contact tip.

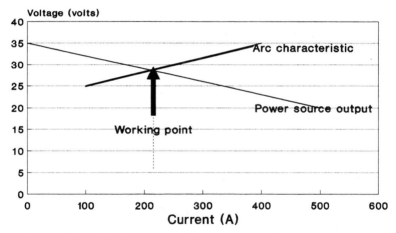

Figure 7.21 Working point with a constant-voltage power source.

In conventional GMAW equipment (e.g. tapped transformer–rectifier designs) voltage is often varied in steps by switches which set the open-circuit voltage and short-circuit current as shown in figure 7.22. The arc voltage must be estimated from a knowledge of the static slope of the power source characteristic. The maximum short-circuit current which is automatically determined

† Stubbing is the undesirable contact of partly melted filler wire with the workpiece.

by selection of the voltage is the maximum level to which the current can rise when the output of the power source is short circuited, for example during dip transfer. With a constant-voltage characteristic power source this level may be very high and result in explosive rupture of the short circuit with high levels of spatter. In order to limit this current it is usual to incorporate inductance in the output of the power source, this reduces the rate of rise of current and limits the maximum value reached before the short circuit ruptures. If the inductance is too high the current will not reach a sufficiently high level to cause detachment and irregular operation will result. In the dip transfer mode the voltage and inductance should therefore be adjusted to optimize the short-circuit current and obtain stable spatter-free detachment.

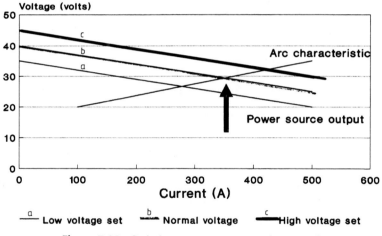

Figure 7.22   Switch output power supply control.

7.5.5   Alternative control techniques

If a constant-current output power source is used the heating effect (current) of the arc does not vary with small changes in stand-off and voltage. Furthermore an inherent self-adjustment mechanism operates with resistive wires since if current and wire feed speed are fixed a unique value of l, the electrical stick-out is defined by equation (7.7). If l decreases momentarily due to some process disturbance the second term in the equation decreases and the melting rate decreases thereby restoring the original extension length. Unfortunately this self-regulation does not occur with high-conductivity materials such as aluminium.

Constant-current power sources may be used for GMAW of high-conductivity materials by using a variable-speed wire feed unit which responds to arc length changes by adjusting the wire feed speed. Electronic AVC (arc voltage control) systems may be used with more modern power sources to cope with this situation. Some of the developments in this area are described in section 7.7.

7.6 Summary: Process Control

A well established relationship exists between mean current and wire feed speed but current must be set indirectly in conventional constant-voltage GMAW systems. The arc voltage, operating current and maximum short-circuit current are normally determined by the open-circuit voltage setting. The rate of rise of current and its peak value during a short circuit are controlled by secondary inductance. Constant-current power sources may be used with resistive wires but additional regulation is required to cope with process fluctuations with high-conductivity consumables.

7.7 Recent Developments in the GMAW Process

The object of the developments of the GMAW process has been to control metal transfer, improve process stability, simplify process control, and improve operating tolerances. The introduction of solid state power sources has enabled the process performance to be analysed in more detail and improved control systems to be developed.

7.7.1 Controlled transfer techniques

The 'natural' modes of transfer which have been described above have several limitations, these include:

(i) Spray transfer only occurs when the mean current exceeds a relatively high transition current. This limits the capability of the process for positional work or the joining of thinner sections.

(ii) Dip transfer, whilst particularly suitable for joining thin-section plain carbon steel, is less effective on non-ferrous materials.

Advances in the control of metal transfer have led to the development of pulsed GMAW and controlled dip transfer.

(a) *Pulsed GMAW*

Pulsed transfer was devised to allow spray-type transfer to be obtained at mean currents below the normal transition level. A low background current (e.g. 50–80 A) is supplied to maintain the arc, and droplet detachment is 'forced' by the application of a high-current pulse (figure 7.23). The pulse of current generates very high electromagnetic forces, as would be expected from the foregoing analysis of metal transfer, and the metal filament supporting the droplet is constricted, the droplet is detached and projected across the arc gap.

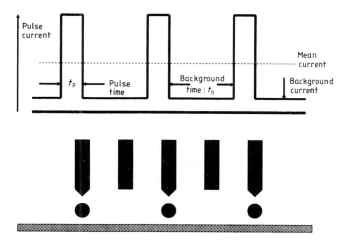

Droplets detached during pulse

Figure 7.23   Pulsed transfer.

The use of high-current pulses to control detachment and transfer metal droplets has been known for some time, but the potential process benefits of low-current spray transfer, with minimum spatter, and improved positional control were not widely realized due to the limitations of early pulsed GMAW equipment. The major problem with early pulsed GMAW equipment was its fixed frequency (50/100 Hz) operation and the additional complexity of establishing and setting the welding parameters. Using the electronic power sources described in Chapter 3 it is possible to generate variable-frequency output waveforms and to optimize metal transfer conditions.

(b) *Controlled droplet detachment*

The pulse parameter required to detach a single metal droplet of a fixed size may be determined for any combination of wire size, composition and shielding gas. The pulse time t_p and pulse current I_p have been shown to follow a relationship

$$I_p^n t_p T = D \tag{7.9}$$

where D is a constant (the detachment constant) and n has a value that is normally between 1.1 and 2. The value of I_p is always above the spray transition current and I_b, the background current, has little influence on detachment. The experimentally determined values are usually plotted in the form of an I_p/t_p curve as shown in figure 7.24. There is some latitude in these experimentally determined pulse parameters for single-droplet detachment, this gives rise to the band of conditions and it is possible to select a range of values which lie within the operating area.

Typical values for I_p and t_p for a range of consumables are given in table 7.3. Ideally the parameters are chosen to produce droplets of similar diameter to that of the filler wire and comparable in size to those found in the drop spray transfer mode.

If it is assumed that a constant length of wire is burnt off at each pulse a simple relationship between pulse frequency and wire feed speed may be defined

$$W = F l_d \tag{7.10}$$

where W is wire feed speed, F is pulse frequency and l_d is the length of wire detached per pulse.

The mean current for a rectangular waveform will be given by

$$I_m = \frac{I_p t_p + I_b t_b}{t_p + t_b} \tag{7.11}$$

158 Advanced Welding Processes

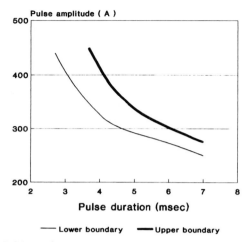

Figure 7.24 Pulse current–pulse duration relationship for one drop per pulse transfer (1.2 mm diameter carbon steel wire in argon/5% CO_2).

where I_m is the mean current and t_b is the background time. Since detachment is controlled by the pulse amplitude and duration the mean current may be reduced to well below the transition current simply by increasing the background time or reducing the pulse frequency. In practice controlled drop spray transfer may be achieved at currents down to 50 A with a 1.2 mm diameter steel wire (transition current $\approx$ 240 A).

Table 7.3 Typical pulse parameters for controlled droplet detachment.

Wire dia.mm	Pulse amplitude I_p(amps) and duration t_p(msec) for a range of common materials					
	Plain carbon steel		Austenitic stainless steel		Aluminium alloys	
	I_p	t_p	I_p	t_p	I_p	t_p
0.8	300	1.5	300	1.5		
1.0	300	2.0	350	2.0		
1.2	350	4.0	350	3.0	250	2.5
1.6	400	4.0			200	5.0

Using the simple equations above it is possible to predetermine the operating parameters for the process as shown in figure 7.25. The procedure is as follows:

(i) Select the pulse current amplitude I_p and duration t_p from the relevant detachment curve or table.

(ii) Choose a suitable mean current (I_m) for the application.

(iii) Determine the required wire feed speed (W) from the burn-off curve for the chosen wire.

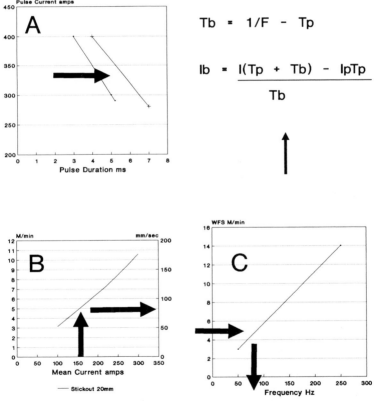

Figure 7.25 Pulse parameter prediction for 1.2 mm wire in argon/5% CO_2. A: pulse parameters. B: burn-off rate. C: frequency versus wire feed speed.

(iv) Select the required pulse frequency (F) from the wire feed speed—frequency curve.

(v) Determine t_b from

$$t_b = (1/F) - t_p \qquad (7.12)$$

(vi) Determine I_b from

$$I_b = [I_m(t_p + t_b) - I_p t_p]/t_b. \qquad (7.13)$$

This procedure allows the operating parameters to be accurately predetermined but it is time consuming and tedious. Computer programs have been devised to overcome the calculation problems but development of power sources with in-built microprocessor control, capable of storing the basic parametric data and the equations above have greatly simplified the selection and setting of suitable welding conditions. Using these techniques further improvements in the control of the process have also been made as described below.

(c)   The influence of pulse parameters on burn-off behaviour

Although it may be assumed that to a first approximation the pulse amplitude and duration may be varied within the limits of equation (7.8) without any change in burn-off behaviour it can be shown that at high pulse currents significant increases in melting rate may occur. This effect results from increased resistive heating during the pulse and the melting rate equation (7.7) may be modified [113] to include this effect

$$\mathrm{MR}_p = \alpha I_m + \beta l \left(I_m^2 + \frac{(I_p - I_b)^2 t_p t_b}{(t_p + t_b)} \right). \qquad (7.14)$$

The significance of the difference between the background and pulse current can clearly be identified in this equation. An increase in the excess current I_e where $I_e = I_p - I_b$ will cause a significant increase in melting rate. Calculated and experimentally determined values for a steel wire are shown in figure 7.26.

In practical welding situations the increased burn-off resulting from the use of high pulse currents may result in weld bead convexity or decreased dilution at a given mean current. It has also been demonstrated [114] that the burn-off behaviour is influenced in a similar manner by variations in the rate of change of current;

rapid rise rates give higher burn-off rates than slower current increases at the same mean current as shown in figure 7.26. For a 1.2 mm diameter plain carbon steel wire the burn-off rate increases by 10% when the slew rate of the current changes from 100 A ms^{-1} to 425 A ms^{-1} The burn-off rate equation may be modified to reflect this effect and equation (7.14) becomes

$$\text{MR}_p = \alpha I_m + \beta l \left(I_m^2 + \frac{(I_p - I_b)^2 t_p t_b}{(t_p + t_b)^2} - \frac{(I_p - I_b)^3}{3S(t_p + t_b)} \right) \quad (7.15)$$

where S is the slew rate $= dI/dt$.

These variations may account for the differences in performance between power sources of a similar design but different dynamic characteristics.

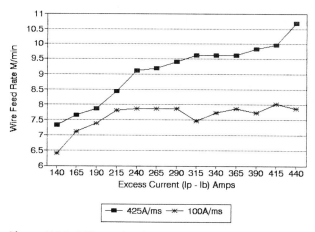

Figure 7.26 Effect of pulse parameters on melting rate.

(d) *Pulsed flux-cored wire welding*

Limited success was achieved using conventional pulsed transfer techniques for flux-cored wire welding but the use of the more flexible variable-waveform power sources described in Chapter 3 has enabled controlled transfer to be obtained with all types of gas-shielded flux-cored wires. The effect of pulsing varies with the wire type but the following phenomena have been identified [115]:

(i) Metal-cored wires behave in a very similar manner to solid consumables and pulsing may be used to enhance positional performance and control whilst at the same time exploiting the alloying potential of the wire. If the appropriate parameters are selected, as discussed in the previous section, the burn-off rate may also be increased at a given mean current.

(ii) Rutile flux-cored wires require slightly different parameters to achieve controlled transfer but these are readily obtained with most commercial advanced pulsed GMAW systems. With the exception of controlling burn-off behaviour the use of pulsed techniques is not easy to justify with this type of consumable since the performance is usually excellent under normal DC operating conditions.

(iii) Basic flux-cored wires normally have poor positional performance and transfer usually occurs in a globular manner. It has been found, however [116], that pulse techniques may be used to improve transfer and enhance positional performance. In the case of basic wires some form of arc voltage control, as described below, is essential.

(e) *Controlled dip transfer*

As discussed above dip transfer is a statistically variable process. The amount of wire transferred by each short circuit is not pre-determined and this will influence arc time and the arc heating available. Attempts to impose a pulsed waveform at a fixed frequency on the process have been unsuccessful because of variations in the amount of metal transferred. In order to achieve an improvement in transfer it is therefore necessary to alter the power source output dynamically to match the rate of transfer.

In one of the earliest attempts at controlled dip transfer a transistor series regulator power source was used to supply controlled pulses of current during the short circuit [117]. A preset level of short-circuit current was initiated at the onset of the dip and current was reduced to the arc level immediately before rupture. The onset of a short circuit was detected by measuring the transient voltage (this falls to zero when the wire comes into contact with the weld pool). Resistance heating of the wire extension immediately after the short circuit has occurred causes an increase in voltage during the short circuit and a sharp rise immediately before rupture. It was found that a short delay was required before the current was increased to allow reasonable contact to be established between the wire and the weld pool. For a 1.2 mm steel wire the duration of this delay was 500 μs after the onset of short circuit. Similarly at the end of the short circuit it was found that the detection voltage should be set to a low enough value to allow the high current to be reduced more than 50 μs before rupture of the short circuit. The resultant process was much more regular than standard dip transfer and spatter levels were much lower. It should be noted that in this case the short-circuit time is not fixed but adapts to the process requirements.

Practical application of this technique was, however, restricted by the sensitivity of the voltage detection level required to sense the imminent rupture of the short circuit. The optimum value may be changed by normal voltage variations, due to changes in stand-off for example. The technique did, however, establish some important requirements for process improvements in dip transfer and based on this work it may be concluded that:

(i) The ideal short-circuit current I_d, short-circuit time t_d relationship are of the form

$$I_d^n t_d = S \qquad (7.16)$$

where I_d is the short-circuit current, t is short-circuit time, n is normally 2, and S is a constant.

(ii) The current at short-circuit rupture should be low to avoid spatter.

(iii) A short delay is required after short circuiting has taken place and before the current increases, to ensure that reasonable contact has been established between the wire and the weld pool.

These 'ideal' requirements are summarized in figure 7.27.

Using the information above it is possible to design a practicable controlled dip transfer system which relies only on the detection of the onset of a short circuit and the re-establishment of the arc. In such systems the maximum short-circuit current is preset but the level of current reached before re-establishment of the arc is controlled by the rate of rise of current. The rate of current rise may be controlled electronically in two stages as shown in figure 7.28. The short-circuit time and peak current will increase or decrease to suit minor fluctuations in the process although the aim is to set the

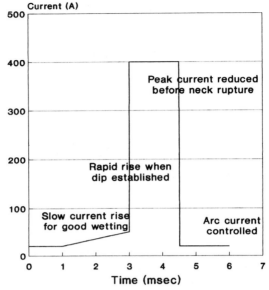

Figure 7.27 Idealized dip transfer waveform (for a single short circuit).

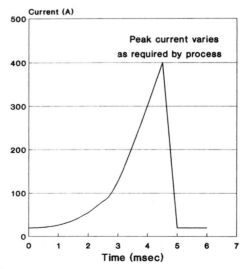

Figure 7.28 Electronically controlled current rise.

parameters so that the majority of short circuits will obey the relationship given by equation (7.12) and the amount of variation is low. The arc current may also be controlled electronically to assist the regularity of the process, improve heat input or force the short circuiting.

7.7.2 Single-knob and programmed control

The complexity of setting welding parameters in conventional DC and pulsed GMAW promoted the development of equipment with 'single-knob' controls as early as 1968 [118,119]. These systems relied on selection of combinations of preset welding parameters (e.g. wire feed speed/mean current and voltage) by means of a single control. They were unfortunately hampered by the power source technology available at the time and in particular by the difficulty of protecting non-electronic designs from mains voltage fluctuations. There was also the problem of producing a limited number of 'ideal' welding parameters which would suit a wide range of applications. Many of these problems were overcome with the introduction of electronic power regulation and micro-processor control, and programmable equipment which can supply a large number of predetermined welding conditions as well as allowing users to record and retrieve their own customized parameters is now available.

A further step has been to incorporate the algorithms described above into control systems which allow continuous control of output over a wide range by the adjustment of a single control. This technique which was originally developed for pulsed GMAW is called *synergic control*.

(a) *Synergic control*

Although in the pulsed GMAW process the optimum welding parameters may be accurately predetermined using the procedures outlined above, if a change in mean current is required the control settings must be recalculated and a number of the welding parameters reset. This could impose significant practical problems, including the possibility of error and resultant deterioration in operating performance. Fortunately it is possible to store both the predetermined parameters and the control equations in the equipment and automatically adjust the output in response to a single input signal. This system is known as *synergic control* [120] and has been defined [121] as follows:

> Synergic control embraces any system (open or closed loop) by which a significant pulse parameter (or the corresponding wire feed speed) is amended such that an equilibrium condition is maintained over a range of wire feed speeds (or average current levels).

In a typical synergic control system the pulse duration and amplitude for single-drop detachment (derived experimentally or from equation (7.9)) are preset. The system may incorporate a tachometer which measures wire feed speed and feeds the speed signal to a control circuit which generates the appropriate pulse frequency. This ensures that a balance between wire feed and melting rate is automatically maintained (using equation (7.10)). When the wire feed speed is varied, either intentionally or accidentally, the welding condition is adjusted to maintain stability. The mean current is determined by the pulse parameters (equation (7.11)). The system is illustrated schematically in figure 7.29.

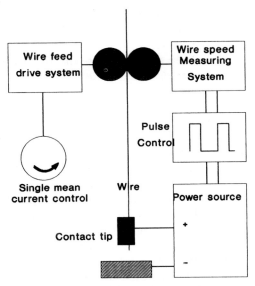

Figure 7.29 Synergic control.

The major advantage of this control technique is that the mean current can be varied continuously over a wide range by means of a single control (e.g. 50–300 A with a 1.2 mm diameter carbon steel wire) and stable projected drop spray type transfer is maintained throughout the control range. Many commercial systems now dispense with the direct wire feed speed feed-back, relying instead on the inherent accuracy and stability of electronic power and wire feed speed regulation.

(b) Developments of synergic control

The synergic control technique has been further enhanced by:

    (i) improved control strategies;
    (ii) arc length control;
    (iii) parameter selection by one user;
    (iv) synergic control of dip and DC spray transfer.

Control strategies. Although these systems should only require a single adjustment it is quite common to incorporate a trim control to accommodate minor deviations in arc behaviour. This fine adjustment usually adjusts the relationship between the wire feed speed and the pulse parameters, changes the burn-off behaviour and lengthens or shortens the arc. It may be required to allow the operator to select arc conditions appropriate to a specific application or to correct any shortcomings in the control algorithm.

Three basic control strategies are adopted. In the original synergic control systems electronic circuits of the type shown in figure 7.30 were used and the effect of increasing wire feed speed was to increase pulse frequency, increase background current and increase pulse height. The pulse duration remained fixed. This is illustrated in figure 7.31. The system gives satisfactory performance over a reasonably wide current range but the pulse current will eventually exceed the value defined by equation (7.9) ($I_p^n t_p = D$) for single-droplet detachment, and control is lost.

An alternative system maintains fixed pulse and background current amplitude, fixed pulse duration and varies only frequency in response to variations in wire feed speed. In this case the 'ideal' pulse parameters are fixed but as frequency rises the background

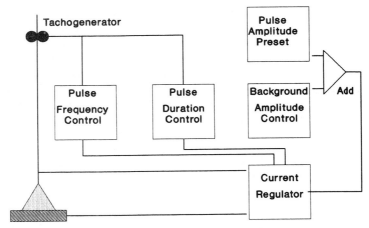

Figure 7.30   Early control arrangement for synergic control.

duration decreases, the preheating of the wire tip during the background period is decreased, and this may reduce the droplet size, which again limits the range of effective control.

The third technique fixes pulse current and amplitude and varies frequency with wire feed speed, but to counter the effects of the variation in background period the product of background

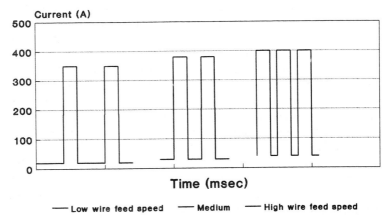

Figure 7.31   Effect of increased mean current and wire feed speed with early synergic systems.

duration and current is maintained constant, i.e.

$$I_b t_b = B. \qquad (7.17)$$

This last control strategy gives an extended range of operation and minimizes the need to trim the conditions as the mean current/wire feed speed is adjusted.

The incorporation of microprocessor control enables more complex control strategies to be adopted, for example it is possible to use a non-linear approach in which the control algorithm is varied with current but it is usually found that the systems described above will give adequate performance in most applications.

Arc length control. The use of constant-voltage power sources in conventional GMAW was justified on the basis of the self-adjustment required. The undesirability of current fluctuation and the need to preset current in pulsed and controlled dip transfer has led to the use of constant-current power sources for these process modes. Under ideal conditions the regular burn-off behaviour achieved with these improved methods of control should ensure constant arc length, but transient fluctuations in wire feed speed, electrical contact in the tip, workpiece surface condition and torch movement may all cause deviations from ideal behaviour. Some inherent regulation of arc length is still obtained with resistive wires, such as steel, operated under constant-current conditions, due to the changes in resistance which accompany any change in electrode extension. For a fixed current and wire feed speed there is a unique (equilibrium) value of stick-out length defined by equation (7.7). An increase in extension causes an increase in resistance which in turn increases the burn-off rate and returns the extension to its original length. Unfortunately since the contribution of the resistance term to the melting rate is negligible for non-resistive wires such as aluminium this mechanism does not operate and transient oscillations in arc length may cause short circuiting or excessive arc lengths. In order to overcome this problem various dynamic control approaches have been adopted. These are usually based on the measurement of the voltage drop in the system and are called *arc voltage control* or AVC systems.

With a constant-current power supply the value of voltage will vary with arc length, even for non-resistive wires, and the melting rate may be modified to correct the change by altering one of the following: wire feed speed, pulse frequency, or pulse height or duration.

The adjustment of wire feed speed can be effective but due to the mechanical inertia in the feeding system its response rate is slow and may be prone to overcompensation.

Pulse frequency can be adjusted within one pulse cycle with more precision than wire feed speed and is used in many systems. In some systems the background current is increased in synchronization with the pulse frequency in order to achieve more effective control.

Pulse height variation in response to arc voltage may be achieved by using an electronically generated constant-voltage characteristic during the pulse and a constant current during the background period. This technique utilizes the same self-adjustment mechanism as conventional systems described above. In this mode the best conditions are usually obtained by using the high-current short-duration parameters defined by equation (7.8) since these are less sensitive to current variation (figure 7.32). Some variation of mean current is inevitable using these systems but since the constant-voltage period only represents a small proportion of the total pulse cycle the current will remain relatively constant at low mean currents.

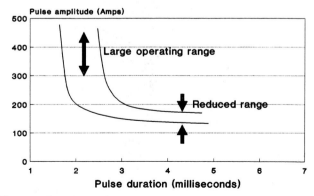

Figure 7.32 Operating range for a pulsed GMAW power source.

Parameter selection by the user. Conventional GMAW equipment may be equipped with electrical, electronic or computer-based systems for storage of from 5 to 100 user-defined parameter sets as discussed in Chapter 3.

In early synergic GMAW equipment the pulse parameters and control algorithms were preset by the manufacturer. This limited the range of applications to those originally programmed and did not allow for new materials or special application requirements. Later units contained the program data in EPROM memory and the equipment manufacturer was able to customize the power source to meet the user's requirements. The latest development in this area has been to provide the user with some programming facility. For example it is possible in some units for the user to set the pulse current and duration for a specific material then use the in-built

Figure 7.33 A hand-held microprocessor programming unit for loading welding parameters to the power source (courtesy GEC).

synergic control strategy to allow automatic adjustment over a wide current range. Another system enables the user to predetermine ideal pulse parameters for a 'new' material using a hand-held computer (figure 7.33) then transfer these parameters to the equipment via a data transfer link. The special user-designed program is stored in non-volatile RAM and may be amended or recalled at any time.

(c) Synergic control of dip and spray transfer

Early one-knob GMAW systems [122] can be considered to have the same features of synergic control as the more sophisticated systems which have recently been developed, but by extending the concepts described above to DC operation with conventional dip and spray transfer it has been found to be possible [123] to produce more flexible synergic control systems which operate with any mode of transfer.

(d) Synergic control systems

The techniques of controlled transfer described above and synergic control techniques are incorporated in many current welding power sources. It is necessary to use electronic power regulation

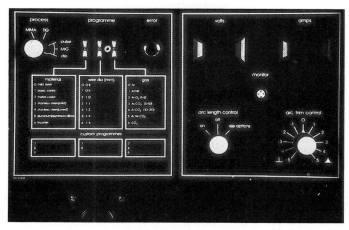

Figure 7.34 Operator control panel of an advanced multiprocess power source (courtesy GEC).

and common to use microprocessor control to achieve the required power source specification. The operator is required to select an initial setting corresponding to the process mode required, the filler wire type, the filler wire diameter and the shielding gas (figure 7.34). The equipment loads the electrical control parameters and control algorithms on the basis of this input information and the only additional adjustment required by the operator is the selection of the mean current required to suit the application.

7.8   Summary

The implications of the GMAW process developments discussed above are that it is possible to improve process reliability and simplify what would otherwise be rather complex control requirements. It is important to recognize that it is still necessary to choose an appropriate metal transfer technique for a given application; whilst the synergic control techniques improve the ease of use of the equipment they do not alter the basic features of the transfer mode.

8 High Energy Density Processes

8.1 Introduction

Advances in the high-energy processes, high-current plasma, electron beam and laser welding, have been concerned with the application of the technology to the fabrication of engineering materials. The power density of these processes is significantly higher than that of the common arc welding processes [124] and normally above 10^9 W m^{-2}; typical values are compared in figure 8.1. As a consequence of the high energy concentration the mechanism of weld pool formation is somewhat different from that normally found in other fusion welding processes. The material in the joint area is heated to very high temperatures and may vaporize, a deep crater or hole is formed immediately under the heat source and a reservoir of molten metal is produced behind the keyhole. As the heat source moves forward the hole is filled with molten metal from the reservoir and this solidifies to form the weld bead. The technique has been called *keyhole* welding.

The high energy density processes which operate in the keyhole mode have several features in common:

(i) they are normally only applied to butt welding situations;
(ii) they require a closed square butt preparation with good fit up;
(iii) the weld bead cross sections have high depth to width ratios (i.e. as narrow gap welds);
(iv) they allow full penetration of a joint to be achieved from one side;
(v) they can be used to limit distortion and thermal damage.

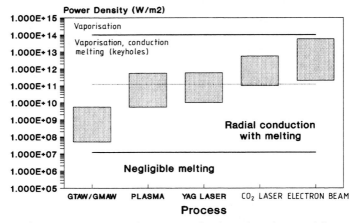

Figure 8.1 Comparative power density of various welding processes. (After Quigley [124] and Ireland [152].)

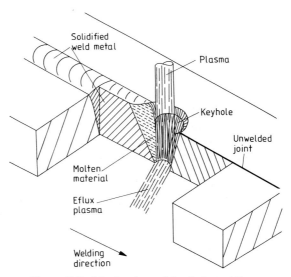

Figure 8.2 Mechanism of keyhole welding.

The mechanism of keyhole welding is illustrated in figure 8.2 and examples of the weld bead profile produced by the technique are given later in this chapter. Very stable operating conditions, and in particular travel speed, are required to maintain a balance between the keyhole-generating forces (gas velocity, vapour pressure and recoil pressure) and the forces tending to close the keyhole (surface tension and gravitational). The need for consistent travel speed as well as safety considerations dictate the requirement to operate these processes automatically.

Although the processes share a common operating mode their individual features and application areas vary and these will be discussed below.

8.2   Plasma Keyhole Welding

The principles of plasma welding have been described in Chapter 6. The same transferred arc operating system is used for plasma keyhole welding but the plasma gas flow and current are usually increased and the orifice size may be reduced. The exact current where the keyhole mode is initiated will depend on the torch geometry and the joint material and thickness but currents of over 200 A and plasma gas flows of $3-4 \, l \, min^{-1}$ are typical with a 2–3 mm diameter constricting orifice. Under these conditions high arc pressures are generated by electromagnetic constriction.

The thermal efficiency of the process is high and it has been estimated that the heat transferred to the workpiece from a 10 kW plasma arc can be as high as 66% of the total process power [125].

8.2.1   Control of plasma keyhole welding

The parametric relationships for plasma keyhole welding are complicated by variations due to torch geometry and it is likely that parameters developed using a specific torch will not be transferable to a torch of a different design. The control variables may be divided as shown in table 8.1 into those normally used to match the conditions to the application, the *primary* controls, and those normally chosen and fixed prior to adjustment of the process, the *secondary* factors.

Table 8.1   Plasma keyhole control factors.

Primary	Secondary
Current	Orifice diameter
Travel speed	Orifice shape/type
Plasma gas flow	Electrode angle
	Electrode set-back
	Shielding and plasma gas
	Torch stand-off

The influence of the main control parameters on process performance may be summarized as follows.

(a) *Mean welding current*

The arc force produced by magnetic constriction is proportional to the square of the mean current. With high currents very high arc forces are generated and undercut or humping may occur (see Chapter 6). The current must, however, be controlled in conjunction with welding speed to produce the required bead profile.

(b) *Plasma gas flow rate*

The presence in the orifice of an insulating layer of non-ionized gas surrounding the plasma is essential. This insulating layer may break down at excessive currents, at low plasma gas flows or when insufficient cooling is available. The resultant instability will often cause double arcing† and serious damage to the torch.

The plasma gas flow rate also influences the arc force and if it is too low the keyhole effect may be lost. At high currents it is therefore necessary to maintain the plasma gas flow rate at a reasonably high level but the upper limit will be determined by the occurrence of undercut and decreased thermal efficiency.

(c) *Welding speed*

If the speed is too high undercutting and incomplete penetration will result, whereas at low speeds the keyhole size may become

† Double arcing is the term used to describe the formation of an arc between the electrode and the orifice and a second arc between the orifice and the workpiece.

excessive and the weld pool will collapse. The operating range of the process is therefore determined by a combination of mean current, welding speed and plasma gas flow. The relationship between thickness penetrated and welding speed at various plasma power† levels have been plotted in figure 8.3. Taking into consideration that this information was obtained from several unconnected sources [126–128] it shows a remarkably consistent trend. Tabulated welding data for plasma keyhole welding of some common engineering materials are provided in Appendix 6.

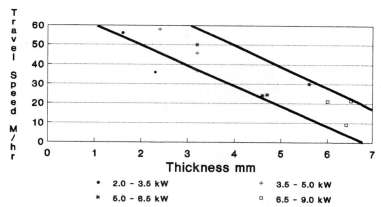

Figure 8.3 Thickness versus travel speed for plasma keyhole welding of stainless steel at various power levels. (Data taken from a range of studies.)

(d) Secondary controls

Orifice diameter. Decreasing the orifice size increases the arc force and voltage. For keyhole welding small, 2–3 mm diameter, orifices are normally used.

Electrode geometry. The electrode geometry and its position within the torch are critical due to their effect on gas flow within

† Plasma power taken as the product of mean current and arc voltage. Voltage is a useful parameter since it takes account of the influence of all the secondary variables on the efficiency of the process. Total arc power therefore gives a good indication of the welding capabilities of the plasma keyhole process.

the torch. It is suggested [129] that tolerances of 0.1–0.2 mm on electrode position are required. Concentricity of the electrode is also important, any misalignment may result in asymmetrical arc behaviour and poor weld bead appearance, the electrode should either be adjustable or fixed by a ceramic insert inside the torch.

Multiport nozzles. The multiport nozzle shown in figure 8.4 may be used to enhance constriction and produce an elliptical arc profile which is elongated along the axis of the weld. Recent work [130] has shown that a second concentric nozzle may also be used to provide an increase in arc pressure although excessive focusing gas flow rates may reduce the thermal efficiency of the process.

Figure 8.4   Multiport plasma welding nozzle.

Shielding gases. The most common shielding and plasma gas is argon. From 1 to 5% hydrogen may be added to the argon shielding gas for welding low-carbon and austenitic stainless steels.† The effect of these small additions of hydrogen is quite significant; giving improved weld bead cleanliness, higher travel speeds and improved constriction of the arc. Helium may be used as a shield-

† Hydrogen additions must be avoided if there is any likelihood of cracking, embrittlement or porosity. Hydrogen should not be used on high-alloy steels, titanium or aluminium alloys.

ing medium for high-conductivity materials such as copper and aluminium. It will tend to increase the total heat input although it may reduce the effect of the constriction and produce a more diffuse heat source. With 30% helium/70% argon shielding gas mixtures keyhole welding speeds 66% higher than those achieved with argon shielding have been reported for aluminium [131]. The shielding gas flow is not normally critical although sufficient gas should be provided to produce effective shielding of the weld area.

(e) *Arc characteristics*

Constriction of any arc causes an increase in arc voltage. In the plasma process, increasing the plasma gas flow, decreasing the diameter of the constriction, increasing current and adding hydrogen or helium to the gas† will all increase the arc voltage. The effect of constriction is shown in figure 8.5. The thermal profile of plasma arcs usually follows a Gaussian distribution as shown in figure 8.6, which also illustrates the effect of shielding gas.

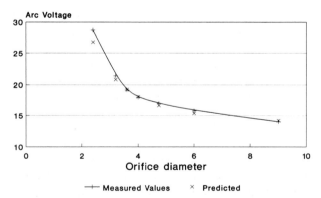

Figure 8.5 Plasma welding: the effect of orifice diameter on voltage. The theoretical equation is $V = 19.55/r^2 + 13.18$ (r is orifice radius).

† Even if hydrogen or helium are only added to the shielding gas there is evidence to show that they will diffuse into the plasma stream and have a marked influence on the process characteristics.

Temperature

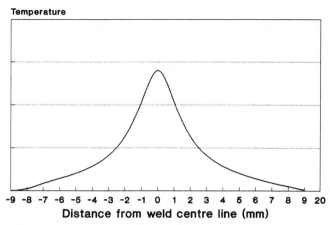

Distance from weld centre line (mm)

Figure 8.6   Temperature distribution of the plasma arc at the workpiece.

(f)   *Reverse polarity plasma*

The plasma process is normally operated with the electrode negative and the workpiece positive. This polarity may be reversed to allow cathodic cleaning to occur when welding aluminium alloys [131]. It is usually necessary to increase the electrode size and limit the maximum current due to the additional heating effect within the torch but helium shielding gas may be used to extend the thickness range weldable in the keyhole mode up to 8 mm.

(g)   *Pulsed keyhole plasma*

The normal keyhole mode of operation is restricted to welding in the downhand position and requires very critical control of speed as discussed above. The tolerance of the process may be improved by modulating the current and positional welds may be made in a wide range of materials and plate thicknesses [132,133]. The current is modulated at relatively low frequencies; the pulse time and amplitude being based on the requirement to establish a keyhole and the background conditions being set to maintain an arc but allow solidification. The resultant weld is therefore formed from a succession of overlapping spots, the travel speed being adjusted to provide at least 60% overlap. Pulsed operation improves the resistance to undercut and generally produces a wider, flatter bead. Examples of typical applications and the features of the DC and pulsed modes of operation are described below.

(h) *Applications*
The keyhole mode of operation makes it possible to perform single-pass square butt welds from one side in plate thicknesses up to about 10 mm.

Carbon–manganese ferritic steel. The plasma keyhole welding of carbon–manganese steels has recently been evaluated for circumferential root runs in pipe for power generation and offshore applications. The use of the process enables thick root sections (6–8 mm) to be welded in a single pass from one side and significantly improves productivity. Pulsed plasma keyhole has been used for these studies [134] to improve operating tolerance and positional performance.

Austenitic stainless steel. Austenitic stainless steel may be readily welded using the keyhole technique and the process has been applied to the longitudinal welding of pipe as well as the fabrication of components for cryogenic service [135]. Welding speeds of around 1.0 m min^{-1} are achievable with keyhole welds in material up to 2.7 mm thick whilst welds in 6.0 mm thick material may be made at 0.35 m min^{-1}. The use of hydrogen additions to the shielding gas or proprietary mixtures containing from 1 to 5% hydrogen provide improved bead appearance and increased travel speed. Undercut may be limited by careful control of welding parameters but if this is not possible pulsed operation or the use of cosmetic passes is recommended.

Nickel alloys. Plasma keyhole welding has been used successfully on a wide range of nickel alloys (including alloy 200, 400, 600, Hastelloy C, Inconel 718) and is particularly useful in the thickness range from 2.5 to 7.5 mm. Argon/5% hydrogen has been used for both plasma and shielding gases for the alloy series 200–600. Typical travel speeds vary from 0.5 m min^{-1} at 160 A with 3 mm thick material to 0.22 m min^{-1} at 310 A with 8.25 mm thick plate.

Titanium. Steady and pulsed keyhole plasma welding techniques have been applied to titanium and its alloys [136,137] and, providing adequate provision is made for gas shielding, high-integrity welds may be made. The shielding requirement may be met by carrying out the welding operation in a glove-box, which is vacuum purged and backfilled with argon, or using a trailing shield of the type shown in figure 8.7. The major problem experienced with this material is undercut; this may be alleviated by:

(i) careful selection of welding parameters on thicknesses up to 3 mm;
(ii) pulsed operation with controlled current decay on each pulse;
(iii) cosmetic runs with plasma or GTAW and filler;
(iv) magnetic arc oscillation along the axis of the weld.

Welding parameters for plasma keyhole welding of titanium are given in Appendix 7.

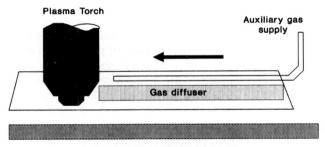

Figure 8.7   A trailing shield.

Aluminium. Plasma keyhole welding of aluminium is possible if electrode positive polarity is used as discussed above. It is also possible to use advanced power sources with the capability of variable-polarity operation and significant improvements in quality and cost have been reported using these techniques [138]. It has been found that acceptable results are achieved with 15–20 ms of DC electrode negative operation and electrode positive pulses of 2–5 ms duration.

8.2.2 Summary—plasma keyhole welding

The plasma keyhole welding process may be used for making square butt welds in a wide range of materials in the thickness range from 2 to 10 mm. It gives high welding speeds and guaranteed penetration from one side of the joint. Although the number of parameters which influence the process performance is large and their interaction is complex, the arc power gives a useful indication of the overall performance capabilities. The major limitation of the process is undercut but this may be controlled by careful selection of the operating parameters or pulsed operation.

8.3 Laser Welding

The laser is a non-arc heat source and consists of a high-energy coherent beam of light of a fixed wavelength. The high-intensity beam is produced by stimulating emission of electromagnetic radiation in suitable gaseous or solid materials.

Helium/neon and CO_2 are commonly used as a basis of gaseous systems whilst ruby and neodymium:yttrium aluminium garnet (Nd:YAG) are used in solid state lasers. In welding applications the two most common types of laser are the CO_2 gas laser and the Nd:YAG solid state laser.

8.3.1 CO_2 lasers

The principles of operation of the CO_2 laser are illustrated in figure 8.8. An electrical discharge within the gas is used to stimulate the emission of radiation. The initial low-level radiation is 'trapped' within the laser cavity by mirrors placed at either end. The internal reflection of the beam causes an increase in the energy level (amplification). A fraction of the laser beam generated in this way is allowed to escape from the resonant cavity via a partially reflective mirror. In the case of the CO_2 laser the emergent light beam will have a wavelength of 10.6 μm (i.e. in the infrared part of the spectrum) and it must be delivered to the workpiece by a series of mirrors and lenses. Since at the operating wavelength glass lenses are unsuitable, lenses for CO_2 systems are usually made from ZnSe whilst mirrors are normally made from copper with a gold reflecting surface. The beam must be delivered to the work-station through rigid tubes and limited beam divergence is essential if long-delivery systems are required. Even with these constraints beam delivery systems have been incorporated into robot welding and cutting installations where flexible positioning of the welding head is required.

CO_2 is the gas which is responsible for the emission of radiation but the mixture in the cavity may contain up to 80% helium and 15% nitrogen. In the process of stimulating emission, heat is generated in the cavity and to avoid overheating and instability the gas is circulated through a cooling system.

The beam power delivered by CO_2 laser systems is commonly between 0.5 and 10 kW and the output may be pulsed or continuous.

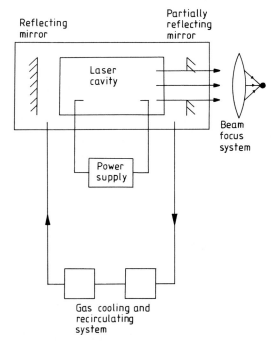

Figure 8.8   Principle of the CO_2 laser.

8.3.2   Nd:YAG lasers

The operating principles of an Nd:YAG laser are shown in figure 8.9. In this case the laser material is made up of a solid yttrium aluminium garnet rod which is doped with neodymium. Stimulation of the electrons in the neodymium is achieved by excitation with high-power flash lamps. Typically the YAG rod has a fully reflective mirror at one end and a partially reflective (approximately 60%) mirror at the beam discharge end. The energy is amplified within the cavity and a beam of radiation in the near infrared range, 1064 nm in wavelength, is emitted through the partially reflecting mirror. Beam delivery can be made much simpler since normal optical glass can be used at YAG laser wavelengths. The cost of the optical components can be reduced and fibre optic cables may be used to provide efficient, flexible distribution systems.

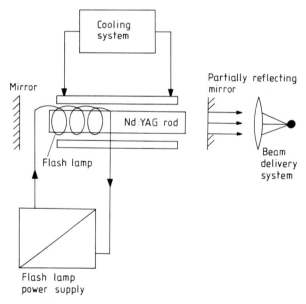

Figure 8.9 Principle of the Nd : YAG laser.

Fibre delivery offers considerable advantages compared with mirror systems; for example they reduce the need for accurate mirror alignment, allow safe delivery of the laser without bulky enclosures and enable variations in laser-to-workpiece distance to be more easily accommodated.

Heating of the rod will occur whilst the laser is running and some form of cooling is necessary to maintain reliable operation.

The range of beam powers available with YAG systems are usually lower than those achieved with CO_2 lasers and for welding 0.1–1 kW average power devices are normally used. The YAG laser can, however, be pulsed to very high peak power levels.

8.3.3   Laser beam characteristics

In operation the beam is focused to a very small spot (e.g. 0.5 mm) at the workpiece surface, the energy distribution across this spot is not uniform but normally has a Gaussian energy distribution in a plane perpendicular to its axis as shown in figure 8.10, this is usually referred to as the *transverse electromagnetic mode$_{00}$* or TEM$_{00}$. The difficulty of defining the outer edge of the Gaussian distribution has led to the convention of measuring the beam diameter in terms of the distance across the centre of the beam in which the irradiance equals $1/e^2$ (0.135) of the maximum irradiance, the area of a circle of this diameter will contain 86.5% of the total beam energy. More complex doughnut-shaped energy distributions may occur as shown in figure 8.10 or a combination of beam profiles may be produced in *multimode* operation where increased output power is required at the expense of beam coherence.

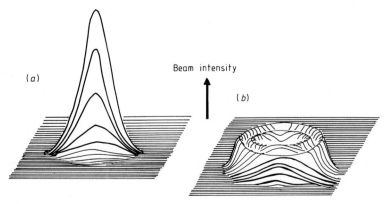

Figure 8.10   Laser beam intensity distribution: (a) TEM$_{00}$ mode; (b) low-order mode.

8.3.4 Welding with lasers

(a) *Welding modes*
Lasers may be used for both melt-in (conduction-limited) welding
in a similar manner to GTAW, or in the keyhole mode described
above. The beam energy delivered to the workpiece will be dissipated by reflection and absorption. In the case of laser welding the
reflected energy losses may be large, many materials being capable
of reflecting up to 90% † of the incident beam energy. The reflectivity is reduced as the temperature of the surface is increased and
if surface heating does occur some energy will be absorbed. The
energy which is absorbed will be conducted away from the metal
surface and if sufficient energy is available to establish a weld

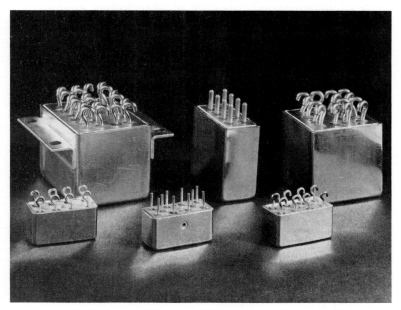

Figure 8.11 Typical   microwelding   application   of   an   Nd : YAG
laser—high-speed seam welding of relay cans with low heat input. (Courtesy of Lumonics.)

† The reflectivity of austenitic stainless steel to infrared radiation
(1060 nm) is 92% whilst that of copper is 98%.

pool, convection within the pool may assist energy transfer. The conduction-limited mode may be used for microwelding applications with low-power (normally below 0.5 kW) lasers; typical applications are shown in figure 8.11.

If a crater can be formed in the weld pool the reflection losses are substantially reduced, the beam energy is absorbed more efficiently, vaporization of the metal occurs and a keyhole is formed. The process may be used with single high-energy pulses to produce spot welds or with continuous or repeated pulsing to produce butt welded seams as discussed above.

(b) *Shielding gases*
It is common to supply a shielding gas to the weld area to protect the molten and solidifying metal. Due to the high travel speeds involved this often takes the form of an elongated shroud which trails behind the beam as shown in figure 8.12. It is also necessary to provide backing gas to the rear of the joint in order to obtain clean penetration beads with satisfactory profiles. The most common gas used for shielding is helium which is inert and due to its high ionization potential is more resistant to plasma formation. Argon is also inert and due to its density should offer improved

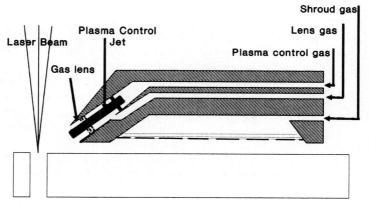

Figure 8.12 Laser welding trailing shield with plasma control jet. (After Hill M *et al* 1990 10 kW CO_2 laser welding of a high-yield-strength steel 2nd Int. Conf. on Power Beam Technology (Stratford-upon-Avon, 23–26 September 1990) (Courtesy AEA Culham).)

shielding efficiency, but some additional plasma control measures may be necessary as a result of its lower ionization potential. The effect of shielding gases on welding speed and depth of penetration is shown in figure 8.13.

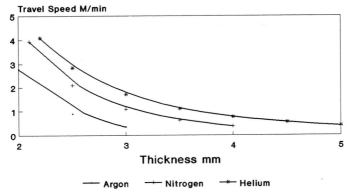

Figure 8.13 Thickness versus travel speed for 2 kW CO_2 laser keyhole welding in various gases.

(c) *Plasma formation*

Interaction between the laser beam, metal vapour and the shielding atmosphere can lead to generation of a visible plasma above the plate surface. The plasma is initiated when free electrons in the surface of the material are accelerated due to the *inverse bremsstrahlung* effect. The accelerated electrons eventually develop sufficient kinetic energy to ionize metal vapour and gases in the area immediately above the plate surface and a plasma is formed.

It has been shown [139] that there is a threshold intensity below which plasma formation is unlikely; this threshold depends on parameters such as beam power, pulse width and the wavelength of the radiation. Typical threshold intensity values for a CO_2 laser are between 1×10^6 and 3×10^6 W cm^{-2} whilst for YAG lasers intensities of 10^8 W cm^{-2} are required. In the initial phases of keyhole crater formation the plasma may assist energy transfer by absorbing energy from the beam and emitting lower-wavelength radiation which is more readily absorbed. In addition the recoil pressure generated by the plasma may assist crater formation. In fact a pronounced improvement in coupling may be observed as the intensity is increased through the level at which plasma formation commences. Once the keyhole has formed, however, the presence of a vapour or plasma plume above the plate surface may limit energy transfer into the plate.

This problem of plasma plume formation after initiation of the keyhole is particularly noticeable with continuous-wave CO_2 lasers operating in the far-infrared region. Although it has also been reported with YAG lasers [140] the plasma formed with titanium and aluminium workpieces under pulsed YAG beams was found to have very low levels of ionization and only resulted in limited loss of power by scattering from metal and oxide particulate.

Plasma control. In CO_2 lasers significant losses of beam power are caused by absorption and scattering from the plasma plume and this has led to the development of the following plasma control techniques:

(i) *Plasma control jets*: The plasma may be deflected or even blown into the keyhole cavity by an auxiliary gas jet [141]. Ideally the jet should be directed toward the plasma to impinge on the workpiece 1 mm ahead of the beam and at an angle of about 20° to the plate surface. A coaxial shield may also be necessary to prevent atmospheric entrainment by the high velocity jet. The arrangement of a typical shield incorporating a plasma control jet is shown in figure 8.12. Helium is again the preferred jet gas due to its high ionization potential but argon and nitrogen have also been used.

(ii) *High-frequency pulsing*: Pulsing of the laser power at frequencies higher than 1 kHz have been shown to be effective in reducing plasma formation in CO_2 laser welding [142]. Both mechanical and electrical oscillation have been used and increased welding speeds and higher depth to width ratios have been obtained.

(iii) *Beam oscillation*: Linear oscillation of the beam along the seam [143] combined with a plasma control jet increases penetration for the same beam power. The gas jet deflects the plasma whilst the linear motion of the beam improves beam interaction.

(d) *Pulsing of laser output*

The output of both CO_2 and YAG lasers may be pulsed to a high level to achieve increased power output. A typical 5 kW CO_2 laser has a continuous-wave power output of 5 kW, pulse frequencies variable between 0 and 25 kHz, pulse width down to 20 μs and peak pulse power of five times the continuous output.† YAG systems with 400 W average power are capable of delivering peak power levels of 5–20 kW for short-duration pulses.

(e) *Process control*

The parameters which control laser welding may be classified as primary and secondary variables as shown in table 8.2.

Table 8.2

Primary variables	Secondary variables
Beam power	Pulse parameters
Travel speed	Plasma control
Focus point	Shielding gases
	Beam mode

Primary controls. The relationship between beam power, welding speed and material thickness is common to most materials and laser types and is illustrated in figure 8.14 [144]. The secondary control variables have a more complex effect on welding performance but some attempts have been made to describe operating envelopes which describe the relationship between welding speed, mode and focus position as shown in figure 8.15. Shielding gas flow can have a pronounced effect on process efficiency and with argon in particular there is a possibility of shielding gas plasma formation at critical flow rates.‡

† Rofin-Sinar RS 5000.
‡ Shielding gas plasma plumes are particularly problematic since they persist in the laser path even after absorption has decreased workpiece coupling. Metal vapour plasma will tend to be suppressed when absorption attenuates the beam.

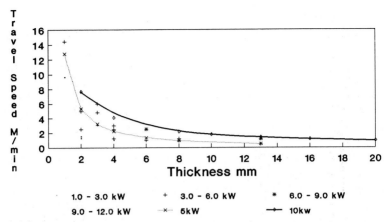

Figure 8.14 Thickness versus travel speed for laser keyhole welding at various power levels and for a range of materials.

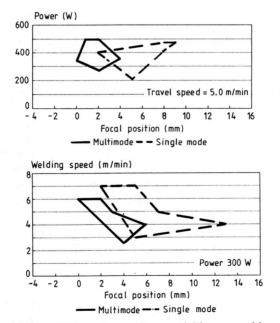

Figure 8.15 Effect of secondary variables on weld quality (for a butt weld by CO_2 laser in 0.15 mm steel). (After Shinmi *et al* [144].)

8.3.5 Applications

The range of materials which can be successfully welded by laser techniques is in the first instance determined by their physical properties, including reflectivity and thermal diffusivity, and secondly by metallurgical considerations. In the keyhole mode once coupling and cavity formation has occurred even metals with high reflectivity (e.g. nickel) may be successfully welded, but copper with high reflectivity and high thermal diffusivity can at present only be welded with difficulty.

Most of the metallurgical problems experienced with laser welding are common to other fusion welding processes; for example, cold cracking in high-carbon and alloy steels, solidification cracking in aluminium alloys, but with suitable precautions these problems may be restricted. Some common application areas are described below.

(a) *Austenitic stainless steel*

Austenitic stainless steels have been laser welded in a range of thicknesses. Typical conditions for 13.3 mm keyhole welds are given in the figures above but speeds of up to 1 m min^{-1} can be achieved with a CO_2 laser at 11 kW.

(b) *Low-carbon steel*

Low-carbon steel is readily joined with a range of common welding processes including GMAW and GTAW. The primary reasons for using high-capital-cost processes such as lasers is to increase productivity and to improve quality. There are no specific problems with the laser welding of uncoated plain carbon steels and both CO_2 and YAG systems have been used successfully as the following applications confirm.

In thin-section sheet material laser welding has been used for fabrication of high-precision pressings [145] to fabricate beams for the carriages for a CNC punch press. A 5 kW laser was used and the main objective was to limit distortion and weld finishing operations.

Lasers are being adopted for many carbon steel welding applications in the automotive industry [146] including the welding of floor panels and engine support frames. In most cases robotic automation is involved and integrated beam delivery systems have been developed. The use of 1 kW YAG lasers with optical fibre delivery systems have also been applied to robotic welding.

Coated steels, particularly zinc-coated or galvanized materials are difficult to weld and even if satisfactory parameters are developed they are prone to batch variation. Some success has, however, been reported using a YAG laser with a multiple laser (Multilase) system [151–152].

Laser welding has also been evaluated for fabrication of thicker-section, higher-strength steels such as ASTM A36 (0.29% C, 0.8–1.2% Mn, 0.15–0.40% Si) [147] and it was found that welding speeds of up to 1 m min^{-1} could be achieved in 19 mm thick plate using a 15 kW CO_2 laser. The cost analysis indicated a three-year payback period for the laser system.

(c)   Titanium alloys
Titanium alloys can be laser welded if due care is taken to prevent atmospheric contamination. Welding speeds up to 6 m min^{-1} have been reported for 3.0 mm thick material using a CO_2 laser at 4.6 kW output power [148].

(d)   Nickel alloys
A range of nickel alloys have been successfully laser welded and welding speeds of 0.5 m min^{-1} have been reported for Inconel 600† when using a continuous-wave CO_2 laser at 11 kW. C263 and Jethete M152 are also readily welded.

8.3.6   Practical considerations

(a)   Joint configurations and accuracy
Square butt, stake/lap, spot, edge, and T butt joints as shown in figure 8.16 are the normal joint configurations used for laser welding. Due to the small focal spot size, alignment of butt joints is critical and joint preparation must be accurate. In addition any automatic positioning, work- and beam-handling equipment must be made to high levels of precision.

† Inconel 600, nominal composition 75% Ni, 15% Cr, 8% Fe.

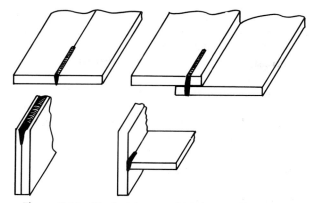

Figure 8.16 Typical laser weld joint configurations.

(b) Safety

Reflection of the beam from the workpiece and surrounding fixtures can be extremely dangerous, particularly with high-power systems, and adequate protective screening must be provided. It is also necessary for appropriate eye protection to be worn.

8.3.7 Developments

(a) Laser-enhanced GTAW and GTAW augmented laser welding

The use of a TIG arc to heat and pre-melt the plate combined with a laser to increase penetration and/or welding speed has been investigated by several workers [149,150]. It has been shown that using a 300 A GTAW arc to augment the laser a 1 kW laser may produce equivalent welding performance to a 2 kW device, and clearly there is potential for extending the range of low-power lasers by this technique. The laser also assists in preventing humping of GTAW welds made at high current and high speeds. Comparisons with multicathode GTAW (see Chapter 6) have not been made but it is likely that similar results could be achieved with this lower-cost process option.

(b) Multiple laser operation

In the case of YAG lasers a novel system has been devised [151] in which three 400 W lasers are brought to a common-output housing by means of fibre optic beam delivery systems. The three lasers may be pulsed in phase to produce the maximum peak output or phaseshifted to give improved control and higher welding speeds.

(c) *Laser development*

Several alternative laser systems are available and improvements in efficiency both in laser generation and application are being developed. Some of the most interesting development areas are:

(i) excimer lasers;
(ii) RF- and microwave-excited CO_2 lasers;
(iii) diode-pumped lasers;
(iv) CO lasers.

Excimer lasers. The active component of the laser medium used for excimer lasers is a rare gas such as xenon, krypton or argon containing a halogen such as fluorine, bromine or chlorine. The pulsed output is in the ultraviolet wavelength range usually from 193 to 350 nm. The average output power of commercial excimer lasers is currently quite limited but strategic target powers of up to 10 kW are forecast [152]. At present the application of this type of laser is restricted to microelectronic production but the photo-ablation process which allows molecular bonds to be broken without introducing excess thermal damage may be used for accurate drilling and cutting.

RF- and microwave-excited CO_2 lasers. DC-excited CO_2 lasers are capable of producing high beam quality at reasonably high power levels. The use of high-frequency excitation has been shown to offer improved beam quality and RF excitation systems for welding are now available. It is expected, however, that microwave systems will offer improved quality, high efficiency and lower overall cost.

Diode-pumped lasers. In general the pumping of solid state lasers by flashlamps is inefficient [153] because of the fairly broad spectrum of wavelengths produced by the lamps and the fairly narrow band of useful pump bands. Greater efficiency can be achieved by pumping with semiconductor lasers such as gallium—

aluminium–arsenide (GaAlAs) which emit wavelengths in the range 750–900 nm. Development of high-power semiconductor-pumped lasers has been limited. Commercial devices with powers of around 1 W are available but the low power limits the applications to areas such as microsoldering.

Carbon monoxide (CO) lasers. Work on the development of CO gas lasers has taken place in Japan [154]. The wavelength of 5 μm falls between that of CO_2 and YAG and may offer potential benefits in the ability to use fibre optic beam delivery. The use of these devices for cutting has been demonstrated but welding applications have yet to be developed.

8.3.8 Summary—laser welding

Laser welding has been shown to be suitable for high-speed welding of thinner materials and deep penetration welding of materials up to about 12 mm in thickness (up to 25 mm is feasible using high-power systems). A wide range of materials is weldable using both CO_2 and YAG systems.

Low-power applications are found in the instrumentation and electronics industries whilst higher-power applications continue to be developed principally in the automotive and aerospace industries.

The capital cost of laser systems is high—£50 000 to £500 000—but the economic returns have justified this level of investment in many applications.

Some of the alternative laser production techniques described above are likely to be developed for welding applications.

8.4 Electron Beam Welding

Electron beams have been used as welding heat sources since the early 1960s and electron beam welding (EBW) has become established as a high-quality precision welding process.

8.4.1 Fundamentals

In the EBW system electrons are generated by passing a low current (e.g. 50–200 mA) through a tungsten filament. The filament is attached to the negative side of a high-voltage power supply (30–150 kV) and electrons are accelerated away from the cathode towards an anode as shown in figure 8.17. The divergent electron beam is focused by magnetic and electrostatic lenses and may be deflected or oscillated magnetically. The complete electron generator assembly or *electron gun* is usually mounted either inside or external to a vacuum chamber which contains the workpiece.

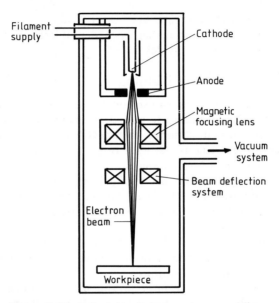

Figure 8.17   Principle of electron beam welding.

8.4.2   Beam characteristics

Power densities of from 10^{10} to 10^{13} $W\,m^{-2}$ are developed at the point of focus and keyhole welding is the normal operating mode.
   The forces which create the keyhole in EBW are:

(i) electron momentum;
(ii) vapour pressure;
(iii) recoil pressure.

Surface tension and gravitational forces counteract keyhole formation but under normal circumstances the keyhole-forming forces are much higher. For example, the electron momentum pressure P_a is given by [155]

$$P_a = 2Jm_eV/e^2 \qquad (8.1)$$

where J is the current density, V the accelerating voltage and m_e is the electronic mass. For a focused spot 0.3 mm radius, 100 mA filament current and 100 kV this force will be around 300 N m^{-2}. The vapour pressure which is temperature dependent can reach values of 5×10^4 N m^{-2} and the recoil pressure at 3.5×10^{10} W m^{-2} has been calculated as 10^7 N m^{-2}. Although it is not possible to equate these forces directly the surface tension force for a 0.5 mm diameter keyhole would be only just over 7×10^3 N m^{-2}.

Under these circumstances very deep penetration keyhole welds can be made with EBW although the general form of the speed/penetration curve (figure 8.18) is similar to that found with both laser and keyhole plasma welds.

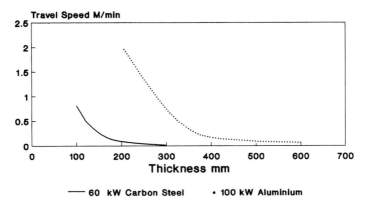

Figure 8.18 Thickness versus travel speed for electron beam welding of various materials.

8.4.3 Control of EBW

Primary and secondary control variables may be identified for electron beam welding as shown in table 8.3. But although the inter-relationship between beam power–travel speed and thickness is clearly established, as shown in figure 8.18, the welding performance may be changed significantly by means of the secondary controls. In particular beam focus and deflection may be used to control the depth-to-width ratio of the welds and intentional defocusing may be used to enable *cosmetic* finishing runs to be made after completion of the penetration weld.

Table 8.3 Control parameters for EBW.

Primary variables	Secondary variables
Filament current	Beam focus
Voltage	Beam deflection
Travel speed	Power supply
	Vacuum

8.4.4 Applications

The EBW process has been used for joining a wide range of materials including alloy steels, nickel alloys, titanium, copper and dissimilar metals in thicknesses ranging from 0.025 mm to 300 mm. Some typical application areas are discussed below.

(a) *Aerospace*

The aircraft engine industry has used EBW extensively for the fabrication of engine parts. A single engine, the Rolls-Royce RB211, utilizes nearly 100 m of electron beam welds [156]. The principal applications include the joining of thick-section stator assemblies in titanium alloys, compressor discs and compressor rotor shafts. The use of EBW has been promoted by the requirement for high-integrity welds with low distortion and minimal thermal damage to the materials.

(b) *Instrumentation, electronic and medical*

The process has been used for the encapsulation of sensors and electronic parts for electronic and medical applications. The materials used include austenitic stainless steels for encapsulation and cobalt–chromium alloys for fabricated hip joints.

(c) *Automotive*

The narrow, deep penetration properties of the electron beam have been used for the circumferential welding of gears to form complex clusters. The process is also used for fabrication of transmission components such as gear cages. Access and high weld quality are primary considerations in these applications as well as the ability to weld *finished* components without distortion or the need for post-weld machining operations.

(d) *Dissimilar metal joints*

The most common production application of EBW to dissimilar metal joints is the butt welding of high-speed steel blade forms to carbon steel backing strip to form hack-saw blades. Although both laser and plasma processes have been used in this application EBW offers very high speeds of up to 10 m min^{-1}.

(e) *Copper alloys*

Unlike laser processes EBW may be used on a wide range of copper alloys and thicknesses up to 12 mm thick may be welded at 0.7 m min^{-1} with beam powers of 10 kW [157].

8.4.5 Practical considerations

(a) *The use of a vacuum*

The EBW process is normally performed in a vacuum to avoid dissipation of the beam by collision with gaseous atoms and to protect both the electron gun and the weld area. The advantage of this technique is that it provides a clean, inert environment which is conducive to the attainment of consistently high joint quality. The major disadvantage is the time wasted in loading and pumping down the enclosure. The harder the vacuum the more difficult it is to achieve, and to alleviate the need for very high vacuum in the complete chamber many systems allow a differential pressure between the gun and the welding area. In these systems the gun vacuum may be maintained at 5×10^{-4} mbar whilst the chamber is held at 5×10^{-2} mbar.

The need to operate in a vacuum also implies the need for cleanliness and the avoidance of low-vapour-pressure compounds in the weld fixturing and positioning equipment.

(b) *Safety*
The collision between the electron beam and a metal surface will generate x-rays; suitable screening is incorporated in the equipment to ensure that the operator is not exposed to this secondary radiation.

(c) *Joint configuration*
The joint configurations are usually variants of square butt, lap and stake welds as previously shown in figure 8.16. Again accurate positioning and joint preparation is usually necessary due to the small spot size.

8.4.6   Developments

Although the basic process has remained unchanged for many years some significant advances have been made in the maximum power available, the operating techniques and the equipment.

(a) *EBW beam power*
Whilst for many applications beam powers of up to 25 kW are quite adequate there has been an attempt to extend the weldable thickness range, particularly for *out of vacuum* applications and systems with output powers up to 200 kW have been built.

(b) *Chamber loading systems*
The chamber loading time may be considerably reduced, particularly for large volumes of small components, by using an integrated loading system incorporating vacuum seals and progressive pressure reduction. The 'shuttle transfer system' developed and patented by Wentgate Dynaweld is illustrated in figure 8.19 and involves the use of custom-designed workpiece carriers into which the component to be welded is loaded. The carrier or shuttle which is fed into the welding chamber via a feed tube are equipped with 'O' rings which seal the enclosure and allow pre-evacuation.

(c) *Vapour shields and beam traps*
The collection of metal vapour inside the electron gun can cause unstable operation, and mechanical shields are often incorporated in the system in an attempt to exclude vapour from the gun area. A further improvement in this area is the development of the

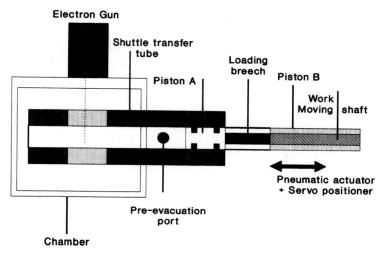

Figure 8.19 Shuttle system for rapid loading of small components. (Courtesy Wentgate Dynaweld.)

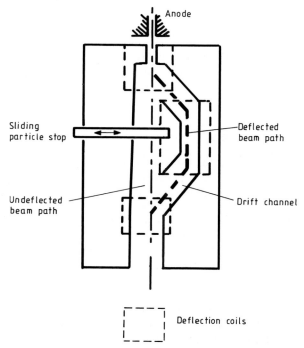

Figure 8.20 The Welding Institute three-bend magnetic trap.

magnetic trap, developed by The Welding Institute and shown in figure 8.20. This system has proved very effective for prevention of gun discharges when welding aluminium alloys [158].

(d) *Computer control*
The incorporation of CNC and computer control is common in current EBW systems. The computer can control workpiece positioning, operating cycle and operating parameters. These systems simplify parameter selection and may often be linked into integrated manufacturing cells. By using telecommunication data links it is also possible to perform remote setting and diagnostics.

(e) *Non-vacuum EBW*
Non-vacuum EBW (NVEB) has become more feasible as beam powers have increased. Production systems are in operation welding up to 6 mm thick aluminium and the development of systems for use in robotic installations is in progress.

8.4.7 Power sources for EBW

The solid state power source technology described in Chapter 3 may be applied to EBW power supplies to provide improved control and consistency. Inverter designs which offer high electrical efficiency are particularly suitable for these applications.

8.4.8 Summary—EBW

Electron beam welding has been used successfully for many years to produce high-quality deep-penetration welds with limited distortion. The advances which have been made in equipment design should enable the productivity of the process to be improved.

Compared with arc welding systems the capital cost of EBW equipment is high but this must be evaluated against the potential improvements in quality and productivity.

8.5 Summary

Three high-energy welding processes have been discussed in this chapter. Of these plasma keyhole welding is well established but

Table 8.4   Comparison of high-energy welding processes.

Feature	PLASMA KEYHOLE	LASER	ELECTRON BEAM
Energy density	3×10^{10}W/m^2	3×10^{11}W/m^2	10^{13}W/m^2
Thickness range	up to 12mm	up to 20mm	up to 200mm
Travel speed	Higher than GTAW	High	Very High
Materials weldable	All, but aluminium needs DCEP or AC.	Most but difficult on reflective surfaces (eg. Cu, Ag)	Most but not low vapour pressure, eg Cd, Zn and coated metal.
System requirements	Simple; as for automated GTAW.	Highly automated plus screening safety interlocks.	Usually vacuum system, X-ray screens and high level of automation.
Capital cost	Slightly more than GTAW.	10 to 30 times GTAW	10 to 30 times GTAW
Running cost	Low	High	Very High
Cost of weld	Depends on application		

probably under-utilized, electron beam techniques are highly developed and exploited in well defined application areas and laser welding is beginning to be applied in a variety of industries. The features of the processes are summarized in table 8.4.

9 Narrow-gap Welding Techniques

9.1 Introduction

Welding process economics may be improved by improved control of the process to give reductions in post-weld inspection and repair, or by decreasing the joint completion time. The aim of many process developments has been to decrease the time taken to complete the joint (so reducing labour costs) by increasing metal deposition rates and using automation.

An alternative approach is to reduce the weld size or joint volume. In fillet welds the possibility of a reduction in weld size will depend on design constraints and the achievement of smaller weld volume is easily controlled by the operating parameters of the process. In butt welds significant changes in weld metal volume may require modification of the joint configuration, a change of process, or both.

The term *narrow-gap welding* is used to describe a group of process developments which have been specifically designed to reduce weld metal volume in butt welds. Most of the development and the application of the processes described below relates to plain carbon and low-alloy steels.

9.2 Principles and Features of Narrow Gap Welding

In conventional 'V' preparations the joint volume, and hence weld completion time, increases dramatically in proportion to the square of thickness (figure 9.1). As the angle of the preparation is reduced the weld metal volume and joint completion rate decrease and if a narrow parallel-sided gap is used the difference becomes significant, particularly on thicker sections. These narrow parallel-sided gaps or square closed-butt preparations are inherent in certain processes, such as flash butt, friction, MIAB, plasma keyhole, laser and electron beam welding. Special techniques must be used, however, to allow the use of narrower gaps with the conventional arc welding processes.

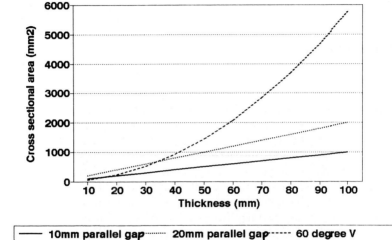

| —— 10mm parallel gap | ⋯⋯ 20mm parallel gap | ----- 60 degree V |

Figure 9.1 The principle of narrow-gap welding—the effect of joint preparation on weld cross sectional area.

In addition to the potential improvement in welding economics which narrow-gap techniques provide they can also give reduced distortion and more uniform joint properties. It has been reported by several investigators [159] that the mechanical properties of narrow-gap joints are better than those achieved with conventional V butt configurations. This is probably due to the progressive refinement of the weld bead by subsequent runs and the relatively low heat input. One practical problem which may arise on thicker sections is the difficulty of post-weld inspection and repair. This leads to the requirement for consistent welding performance and adequate in-process control and monitoring.

9.3   Narrow-gap Welding Processes

The narrow-gap welding processes have certain common features:

(i) they use a special joint configuration;
(ii) they often require a special welding head/equipment;
(iii) they usually require arc length control/seam tracking;
(iv) they may require modified consumables.

The simplest joint configuration is the straight, parallel-sided gap with a backing strip, but there are a number of variants of preparation based on the process and the nature of the application. Some typical preparations which have been used are shown in figure 9.2. The gap width also varies depending on the process and

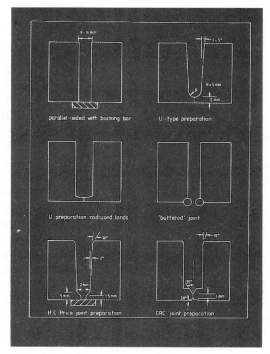

Figure 9.2 Typical narrow-gap joint preparations. (After Foote W 1987 *PhD Thesis* Cranfield Institute of Technology.)

the equipment; from around 8 mm for GTAW up to 20 mm with SAW.

It is possible to use standard automatic welding equipment with some processes but with thicker sections (over 100 mm†) it is necessary to use purpose-designed torches to provide access and ensure adequate gas or flux cover. Standard power sources and wire feed systems may also be used, but it is important that these give a stable and reproducible output. The torch height and its position relative to the sidewalls of the gap must be maintained at predetermined values, and this usually entails some form of seam tracking and height-sensing system. The consumables may require modification to give a satisfactory bead profile and, in the case of slag-shielded processes such as FCAW and SAW, to allow the solidified slag to be easily detached.

9.3.1   GTAW

(a)   *Cold wire GTAW*
The GTAW process is easily adapted for narrow-gap operation since the compact welding torches are available and in thinner sections it is only necessary to extend the distance by which the electrode projects from a conventional gas shroud (figure 9.3). Some provision must also be made for feeding a filler wire into the weld pool, but again conventional equipment is often suitable for thicknesses up to 12 mm. For thicker sections a special torch design is required to ensure adequate gas cover and reliable wire positioning. Torches specifically designed for narrow-gap GTAW are usually elongated as shown in figure 9.4, the shielding gas is delivered to rectangular slots either side of the electrode and in addition gas may be supplied through holes in the side of the blade.

Whilst the design is in principle fairly straightforward it has been found that the achievement of adequate gas cover without entrainment of air from the surroundings is quite difficult [160]. Telescopic gas shrouds and flexible surface baffles are often added to supplement the gas shroud as shown in figure 9.5.

---

† The limiting thickness for changing from conventional to special-purpose narrow-gap equipment is not fixed but depends on the application and the process economics.

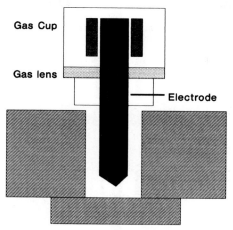

Figure 9.3 Narrow-gap welding with a conventional GTAW torch. (Note that some auxiliary shielding may be needed.)

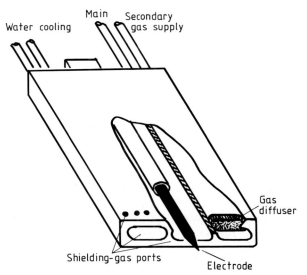

Figure 9.4 Narrow-gap GTAW torch.

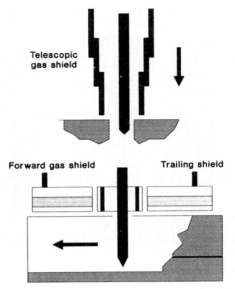

Telescopic
gas shield

Forward gas shield   Trailing shield

Figure 9.5 Additional gas-shielding facilities.

The process has been used for welding thicknesses of up to 200 mm particularly for circumferential welds and pipework.

(b) *Hot wire GTAW*

The hot wire GTAW process described in Chapter 6 is particularly suitable for narrow-gap welding since it allows the deposition rate of the process to be increased to levels comparable with those of GMAW. Many commercial applications of narrow-gap GTAW employ hot wire addition and an interesting example is the Roboweld system shown in figure 9.6 which was designed for welding transmission pipelines both on land and on pipe lay barges. Seam tracking and accurate positioning of the electrode may be achieved by one of the methods described in Chapter 10, but some systems also include facilities for video monitoring of the arc and torch position with the ability for the operator to override the automatic control to correct the torch position. To improve the sidewall fusion the electrode may be oscillated or alternatively the arc may be deflected magnetically.

Figure 9.6 Multihead narrow-gap and hot wire GTAW system for line pipe welding (Saipem Roboweld). (Courtesy EMC UK.)

9.3.2 GMAW

The potential for using narrower weld preparations and smaller included angles for GMAW has often been claimed as an advantage of the process when compared with SMAW. But although some concessions are allowed in the construction standards this advantage has not been fully exploited in manual welding probably due to the difficulty of maintaining consistent fusion with reduced access. Reduced preparation widths are, however, used in automatic welding where improved control is possible. The developments in narrow-gap GMAW fall into two categories:

  (i) the use of conventional automatic systems and narrow*er* gaps;
  (ii) the use of special narrow-gap GMAW systems and narrow gaps.

(a) *Reduced gap/angle* GMAW

A major application in the first category are the systems developed for transmission pipeline welding. The CRC system [161], for example, uses a compound bevel of the type shown in figure 9.2. In comparison with a standard American Petroleum Institute (API) bevel this preparation reduces the weld metal volume by more than 20%. The pipe position is fixed with its longitudinal axis in the horizontal plane and welding must therefore be carried out in the 5G position (i.e. with the welding system rotating around the pipe). This is achieved by using a tractor equipped with an oscillator and typical operating conditions are shown in table 9.1.

Table 9.1   Welding conditions for CRC pipe welding procedure.

Welding variable	Conditions for each pass				
	Joint preparation (Cap, Fill runs 4 3 2 1, Hot pass, Root)				
		Root run	Hot pass	Fill 1-3	Fill 4	Cap
Travel speed		760 mm/min	1000 mm/min	330 mm/min	330±10% mm/min	330±10% mm/min
Wire feed speed		8.6 m/min	12.7 m/min	14.6 m/min	14.6 m/min	9.4 m/min
Shielding gas		75% argon, 25% CO_2	100% CO_2	100% CO_2	100% CO_2	100% CO_2
Gas flow rate		1.4 m^3/hr	1.4 m^3/hr	2.8 m^3/hr	2.8 m^3/hr	1.4 m^3/hr
CTWD*		6.3mm	8.9mm	14-12.7mm	12.7mm	7.9mm
Voltage		19 volts	23.5 volts	23.5 volts	23.0 volts	20.5 volts
Approx. Current		190 amps	250 amps	270 amps	270 amps	200 amps

Note: * CTWD = Contact tip to workpiece distance.
Parameters derived from; Randall.M.D,Nelson.J.W., CRC Automtic Welding System, Paper 2,pp 6-13, Proc. Conf. Recent Developments in Pipeline Welding Practice, Published by The Welding Institute, Cambridge, 1979.

Systems such as this have been used successfully for welding pipe diameters of 600–1500 mm with wall thicknesses of 8–22 mm.

(b) Narrow-gap GMAW developments

The use of narrow-gap GMAW for the welding of submarine hulls was reported as early as 1966 [162] and by the mid 1970s a range of developments of the process were being used on production applications in Japan.

Very thick sections may be joined by GMAW using a parallel-sided gap if special torches are employed; these torches, like narrow-gap GTAW torches, are blade-shaped with elongated gas delivery nozzles. Several different approaches have been adopted in an attempt to minimize the risk of lack of fusion in the narrow-gap GMAW process; these are listed below.

Bead placement. The fusion may be controlled by careful placement of the weld beads and the use of a two pass per layer technique as shown in figure 9.7. The torch is re-positioned after each run or alternatively two separate torches may be used.

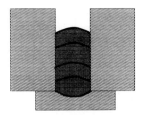

Single pass per layer

Two pass per layer

Figure 9.7 Single and two pass per layer techniques.

Consumable modifications. The fusion characteristics of the process may be improved by using a high level of CO_2 in the shielding gas but this often leads to low process stability and excessive spatter build-up; whilst this may be acceptable when conventional torch designs are used it is likely to cause operational problems with specialized narrow-gap torches. The alternative is to use high helium gas mixtures (helium + argon + CO_2 + oxygen). It has been found that these give superior penetration to argon-rich mixtures and provided the oxygen and CO_2 levels are carefully controlled the arc stability is good [163] and the mechanical properties of the joints are superior to those produced with either CO_2 or argon/CO_2 mixtures.

The position of the filler wire within the joint determines the arc

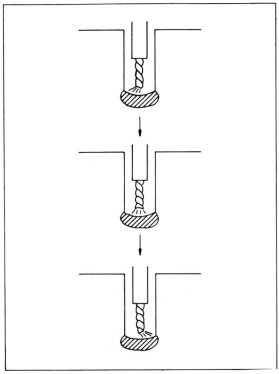

Figure 9.8   'Twist arc' narrow-gap GMAW technique.

root location and will influence the incidence of fusion defects. Normal GMAW filler wires have a natural curvature (referred to as *cast*) and spiral (referred to as *helix*), and this gives rise to random oscillation of the wire tip. Specially straightened wire is available for automatic welding and automatic equipment often incorporates some means of wire straightening. These techniques ensure a stable arc location, but a novel modification of the filler wire, which exploits the presence of a helix to produce more controlled arc oscillation, has been used as a means of improving fusion. This technique, known as *twist arc* involves fabricating a special filler wire by twisting two smaller-diameter wires together [164]. The principle of the technique is shown in figure 9.8. The welding head tracks along the longitudinal axis of the weld and with a gap width of 14 mm adequate fusion is obtained at both sidewalls. The arc oscillation pattern may be varied by changing the relative diameter of the wires, the pitch of the spiral and the operating parameters.

Torch or wire oscillation. Lateral oscillation of the torch may be used as shown in figure 9.9, or alternatively an eccentric contact tube may be rotated to produce circular oscillation of the wire tip as shown in figure 9.10. These systems are relatively complex and the minimum gap width is often limited by the need to move the whole welding head. Alternative systems that rely on controlled deformation of the filler wire as it passes through the feeding system have therefore been developed. These devices use bending rollers to introduce a wave-like or spiral deformation into the wire in order to cause controlled oscillation of the arc. Although the mechanism required to produce the deformation may be complex the part of the torch which enters the joint is compact and satisfactory joints can be made in gap widths down to around 9 mm.

Modification of process operating mode. Many of the systems now available use the pulsed transfer GMAW techniques described in Chapter 7 to improve process control and limit spatter formation.

(c) *Applications of narrow-gap GMAW*
The narrow-gap GMAW process has been used for downhand welding of circumferential joints in pipe and shafts with the workpiece rotated under the welding head. The GMAW process may also

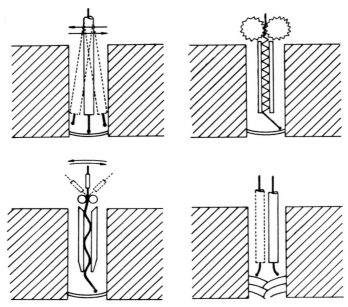

Figure 9.9   Oscillation techniques for narrow-gap GMAW.

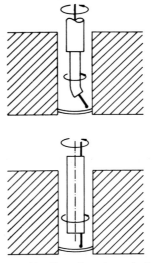

Figure 9.10   Rotating tip techniques.

be applied to positional (5G) welding of pipe as described in section 3.2.1 and has also been used for welding horizontal joints in tubular structures and building columns as shown in figure 9.11, particularly in Japan.

In the offshore industry interest has been shown in the application of the process to horizontal welds (2G) in vertical pipe for 'J' laying of transmission pipelines at sea. Although this has been found to be feasible it requires very careful control of the process parameters and gap width to ensure that consistent bead profile and fusion characteristics are maintained.

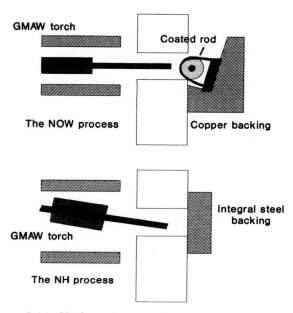

Figure 9.11 Horizontal narrow-gap techniques. (From Nakayama M and Arikana M 1975 Development and application of narrow gap arc welding processes in Japan *Welding in the World* **13** (9/10).)

9.3.3 Narrow-gap SAW

The narrow-gap submerged arc process is capable of producing high-quality joints in thick sections in the downhand position with

considerable improvements in running costs. Conventional submerged arc welding equipment may be used for relatively thick
material (e.g. 70 mm) but special-purpose equipment is available
for welding thicknesses up to 600 mm.

(a)   *The single-pass technique*

The use of a single-pass technique, which is potentially more pro-
ductive, is limited by the possibility of sidewall fusion defects, slag
entrapment and a higher average heat input when compared with
multipass modes. Recent work [165,166] using mathematical
modelling techniques has shown, however, that it is possible to
optimize the welding parameters so that defects such as lack of
fusion, undercut and slag removal problems may be overcome in
thicknesses up to 70 mm with conventional equipment and a single
pass per layer. A microsection of a completed weld in 70 mm plate
is shown in figure 9.12 together with the welding parameters used.
It was found that using the parameters predicted by the model the
bead geometry could be controlled to produce concave surfaces
(for ease of slag detachment), optimum depth-to-width ratios (to
prevent solidification cracking) and maximum lateral penetration
(to prevent sidewall fusion defects).

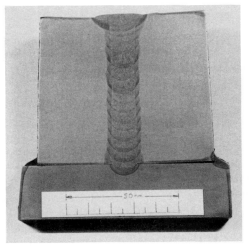

Figure 9.12 Microsection of narrow-gap weld in 70 mm
thick steel using the single run per layer technique.

(b) *The multipass per layer technique*

The multipass technique requires a larger gap width for a comparable wire diameter (e.g. 18 instead of 10 mm with a 4 mm filler wire and two passes) and the joint completion rate is consequently lower. The technique does, however, allow greater control of sidewall fusion, weld metal refinement and improved access for flux removal and interrun cleaning. For these reasons most commercial applications of the process use the two pass per layer mode with a single electrode. Typical values of gap width and welding parameters are given in table 9.2 [167]. For thicknesses above 100 mm a special narrow-gap torch is required; this is usually rectangular with provision for flux feed and recovery, seam tracking and height control. A typical torch assembly is shown in figure 9.13. This particular unit uses tactile sensors to control torch height and optoelectronic sensing of the torch position within the gap. The process is commonly applied to longitudinal and circumferential joints in large cylindrical components; in both cases a heavy column and boom will be required to carry the torch head and in the case of circumferential joints the workpiece must be rotated under the welding head. For these circumferential applications it is essential to prevent fluctuation in the lateral position of the seam as it rotates, and feedback control devices are often incorporated in the roller bed to sense and correct the component position.

Table 9.2 Narrow-gap submerged arc parameters.

Groove width mm	Current amps	Voltage volts	Travel speed mm/min	Flux height mm
12	425	27	180	45
14	500	28	250	50
16	550	29	250	50
18	600	31	250	50
20	625	32	230	50
22	625	34	230	50

After; Hirai.Y.et al.,Application of narrow gap submerged arc welding process to fabrication of 2¼%Cr-1%Mo forged steel heat exchangers. IIW Doc.XIIA-009-81, 1981.

Single filler wires are used with the most commonly used wire diameters being in the range 3.2–4.8 mm in diameter; smaller wires are prone to cause random arc wander, particularly with the relatively long electrical stick-outs which are used, and larger diameter wires are more difficult to feed. Although it is possible to use standard fluxes, special flux formulations with improved slag release characteristics have been devised for narrow-gap SAW. DC electrode positive or AC power is used; AC gives greater resistance to magnetic arc blow and square wave AC power supplies have also been used to improve control of the process.

Figure 9.13   Narrow-gap SAW system. (Courtesy Ansaldo–Italy.)

(c) *Developments*
Increased productivity and improved process control may be achieved by applying the techniques normally applied to conventional submerged arc welding to the narrow-gap process, i.e.:

(i) extended stick-out
(ii) twin wire
(iii) hot wire
(iv) metal powder addition
(v) flux-cored consumables

These techniques are not widely used at present although some development work has been reported [168,169].

(d) *Applications*

The process has been in use in many commercial applications since the early 1980s [170]. Some of these are summarized in table 9.3; they range from nuclear reactor containment vessels in 600 mm thick Ni/Cr/Mo alloy steels to the welding of 60 mm material for offshore tubulars.

Table 9.3 Applications of narrow-gap submerged arc welding.

Weld type	Application
Circumferential rotated pipe.	Offshore oil platform tubular fabrication
Circumferential rotated shafts	Power generation turbine components. Propeller shafts.
Rotated thick wall vessels	Nuclear reactor containment vessels

9.4   Summary and Implications

The use of narrower joint gaps and reduced preparation angles can result in significant improvements in productivity. The use of processes which involve the inherent use of a narrow gap (EBW, laser, plasma, friction) automatically exploit these advantages, whilst systems have been developed to allow narrow gaps to be used with GTAW, GMAW and SAW processes. The potential reduction in running costs must be evaluated against the capital cost of the equipment, although it is reported [171] that sophisticated narrow-gap submerged arc systems costing as much as £250 000 have been justified for welding 350 mm thick high-pressure feedwater heater shells. The *minimum economic thickness* for narrow-gap technology varies with the process and operating mode. Narrow-gap GTAW welding may be justified on thicknesses down to 15 mm. Narrow-gap type configurations have been used for GMAW in thicknesses from 15–22 mm upwards. Narrow-gap SAW is normally considered to be viable at thicknesses above 60–70 mm but if conventional equipment is used this lower economic limit may be reduced until it overlaps conventional square butt SAW procedures in thicknesses down to 12 mm. Optimization of welding parameters and in-process control are essential to avoid defects in narrow-gap applications; the restricted access of the gap will make progressive repair difficult, but good procedure control should obviate these problems and in fact the use of narrow-gap welding may be seen as a way of imposing a reasonable level of discipline into the control of welding operations.

10 Monitoring and Control of Welding Processes

10.1 Introduction

In any manufacturing process it is necessary to ensure that the outcome of the operations carried out matches some defined objective. This involves controlling the operation of the process in some way as shown in figure 10.1.

In order to achieve the desired output it is necessary to:

(i) Establish *control relationships* which enable the effect of the control variables on the process performance to be predicted.

(ii) Monitor the process to ensure that it is operating within limits defined by the control relationships.

Compared with other manufacturing processes welding has established a reputation for being more difficult to control and less likely to achieve consistent quality. This is probably a result of the multiplicity of interrelated control parameters, the complexity of the control relationships and the difficulty of monitoring process performance. An indication of the number of control parameters which need to be considered for two contrasting arc welding processes is given in table 10.1.

The traditional manual control techniques will be outlined in this chapter and the influence of recent developments in processes and equipment will be discussed. Substantial progress has also been made in both the determination of control parameters, monitoring techniques and automatic process control, and these advances will be described in detail.

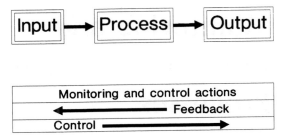

Figure 10.1   Principles of process control.

Table 10.1   Comparison of control variables for automatic plasma and MMA welding.

Control Parameters SMAW	Control Parameters PLASMA
Electrode type Electrode diameter Arc current Direct or alternating current Electrode polarity Electrode manipulation	Current, pulse parameters, current rise/decay times, electrode polarity, welding speed, electrode geometry, shielding gas type, shielding nozzle size, shielding gas flow stabilisation (gas lens), shielding gas flow rate, electrode protrusion. Electrode set back, nozzle geometry, orifice diameter, plasma gas type, plasma gas flow, pilot arc current.

10.2   Manual Control Techniques

Traditionally, welding processes have been controlled by establishing satisfactory operating envelopes for a particular application, often by trial and error, recording the most satisfactory parameters and using these in production. In some cases it has been left to the welder to interpret inadequate drawings and establish conditions which satisfy the design requirements; for example to produce a fillet weld of a given size.

When improved control is required a *welding procedure* is established. This is a formal record of the parameters which have been found to produce the required result and it is used to specify the steps necessary to achieve repeatable weld quality. Procedure control has become the accepted approach when high-quality joints are being produced.

10.2.1   Formal welding procedure control

Formal welding procedure control entails:

(i) establishing satisfactory operating parameters—*procedure development*;

(ii) gaining acceptance of the proposed procedure—*procedure qualification*;

(iii) following the accepted procedure in practice—*procedure management*.

(a)   *Procedure development*

Welding procedure development involves: selection of the most suitable welding process; the determination of a suitable combination of welding parameters; assessment of the performance of sample joints; and amendment of parameters if test results fail to meet requirements. These factors are discussed below.

Selection of welding process. Choice of the process will depend on the material to be joined, its thickness and the welding position. In most cases several processes will meet the basic requirements of the application and the final choice will depend on practical considerations (e.g. availability of equipment and operators), limitations imposed by codes (see below) and economics. The choice of process will determine the number of control parameters which need to be considered and the nature of the control relationships. Computer software designed to simplify welding procedure selection is also available [172].

Determination of welding parameters. The *welding parameters* include all the variables which need to be specified in order to ensure repeatable performance. This may involve the joint design, cleaning and edge preparation, preheating and post-weld treatment as well as the process control parameters such as speed, voltage and current.

The application may call for a specific joint configuration, but it is usually necessary to define the details of the plate preparation. Predetermined joint profiles are available from published literature [173], welding codes and standards (e.g. BS 5135, BS 4145, AWS D1.1). These codes provide 'safe' preparation details which have been tested for the application covered by the standard. They provide a simple method of joint design, but in some cases they may restrict the choice of process and joint profile and it may still be necessary to qualify the procedure if a new process/technique is to be used.

The process control parameters can also be determined by reference to published data, pre-qualified procedures, or codes and standards. Alternatively, welding trials may be necessary to determine suitable parameters, operating tolerances and the optimum welding conditions. At this stage the combination of process and parameters chosen must be capable of producing a joint of the

Table 10.2 Typical examples of the influence of material weldability on procedure.

Material	Problem	Influenced by	Control
Carbon manganese and low alloy steel.	Hydrogen induced cold cracking	Microstructure, stress, restraint	Carbon equivalent of material. Electrode type (Hydrogen controlled)He at input. Cooling rate. Preheat and postheat.
Carbon manganese steels.	Lamellar tearing	Plate cleanliness, presence of through thickness stress.	Check material. Correct joint design. Avoid through thickness loading.
Ferritic steels.	Porosity	Presence of nitrogen (from air). Generation of carbon monoxide.	Use adequate filler deoxidation and efficient gas shield.
Austenitic stainless steels.	Carbide precipitation	Material composition. Thermal cycle.	Coose stabilised material. Control thermal cycle.
Austenitic stainless steels.	Hot cracking in weld metal.	Weld metal composition, heat input.	Control ferrite content (electrode).

WELDING PROCEDURE SPECIFICATION

PROJECT:	W.P.S. No.:	Rev.:	Page:

		APPROVALS
WELDING CODE	:	
WELDING PROCESS	:	CONTRACTOR: ....
EDGE PREPARATION	:	CLIENT: ....
JOINT TYPE	:	CERT. AUTH.: ....
JOINT POSITION	:	
DRAWING REFERENCE	:	

JOINT TOLERANCES

DIMENSION		AS DETAILED
ROOT OPENING	R	....
ROOT FACE	F	
GROOVE ANGLE	OC	....

MATERIAL SPECIFICATION

GRADE: ....
THICKNESS: ....
RANGE QUAL.: ....

THERMAL TREATMENT

PREHEAT: ....
MAX. INTERPASS: ....
P.W.H.T.: ....

JOINT DETAIL   PASS SEQUENCE

WELD PASS DETAILS			ELECTRODE DESCRIPTION			FLUX / GAS	WELDING PARAMETERS				INTER PASS	HEAT INPUT
No.	SIDE	POS	TYPE	SIZE	SPEC.		AMPS	VOLTS	POL	SPEED	TEMP	KJ/mm

TEST PLATE IDENTIFICATION	CONSUMABLE/DETAILS	N.D.E.
MARKED:	STICKOUT:	VISUAL: ....
WELDER:	SHIELD GAS:	M.T.: ....
No.:	FLUX:	R.T.: ....
DATE:	ELECTRODE	U.T.: ....

NOTES:	PREPARED BY: ....

Figure 10.2 Typical welding procedure specification sheet.

specified repeatable quality cost-effectively. The avoidance of potential defects must be taken into account when choosing the process and may significantly influence the selection of welding consumables, operating parameters, pretreatment, post-weld heat treatment and inspection.

The problems which need to be considered are determined by the weldability of the material as indicated in table 10.2. In the case of structural steel the possibility of hydrogen-induced cold cracking (HICC) must be considered. Fortunately the rules governing the formulation of safe procedures to avoid HICC are well established and are covered in most national and international codes (e.g. BS 5135). These rules indicate the need for hydrogen-controlled filler materials, and enable the preheat and heat input requirements to be determined from a knowledge of the chemical composition of the steel or its carbon equivalent and the combined thickness of the material. The result of the development process will be a formal welding procedure specification (WPS) which may consist of a simple list or more usually it will be produced as a printed form (figure 10.2).

Assessment of joint performance. In order to test whether the procedure will produce the required joint characteristics it will be necessary to carry out either a mechanical or non-destructive examination of sample welds which are made with the specified welding parameters.

Amendment of procedure. If the specified procedure fails to produce the results which were required it may be necessary to repeat the process and amend the welding parameters.

(b) *Procedure qualification*

Formal qualification of the procedure involves completion of sample joints to the agreed WPS, often under the supervision of the client or an independent approval body. The welded joints are subjected to a specified selection of non-destructive and mechanical tests, the results of which are reported in a procedure qualification record (PQR) as shown in figure 10.3.

The skill of the welder is often a major factor in determining the final weld quality. The welder's skill may be assessed and '*calibrated*' by means of a general approval process [174]. It may

WELDING PROCEDURE QUALIFICATION

PROJECT	W.P.Q. No.:	Rev.:	Page:

WELDING CODE:		APPROVALS

TESTING CODES

TENSILE:	N.D.E.	CONTRACTOR:
HARDNESS:		CLIENT:
TOUGHNESS:		CERT. AUTH:
BEND:		

NON DESTRUCTIVE EXAMINATION / MACRO EXAMINATION

TEST	RESULTS	REPORT	TEST	RESULTS	REPORT
VISUAL			CONTOUR		
MAG. PARTICLE					
DYE PENETRANT			FILLET SIZE		
ULTRASONIC					
RADIOGRAPHY					

TENSILE TEST

TYPE	TEST No.	AREA	ELONG. %	R of A	YIELD	U.T.S.	COMMENTS	REPORT

HARDNESS SURVEY / BEND TEST

TRAVERSE	PM	HAZ	WM	COMMENTS	REPORT	TYPE	FORMER	RESULTS	REPORT

TOUGHNESS TEST

TYPE	No.	TEMP	RESULT	REPORT

TEST PLATE IDENTIFICATION

MARKED:	NOTES:
WELDER:	
No.:	
DATE:	

Figure 10.3 Typical welding procedure qualification record sheet.

also be required to qualify a named welder to carry out a specific procedure.

(c) *Procedure management*
Management of welding procedures using this technique involves: maintenance of procedure specification and qualification records; calibration of welding and ancillary equipment; and monitoring compliance with the specified procedure.

Maintaining procedure records. Having established a satisfactory procedure and obtained procedure qualification it is necessary to maintain a record of the parameters and techniques used and to control the issue of this information to the shop floor to enable consistent joint quality to be achieved. In many cases this information will be required on an irregular basis particularly when small batches of fabrications are required. The welding procedure specification records will also be required when tendering for new projects, when preparing procedures for new work or when analysing production problems. A fabricator will quickly generate a large library of procedures and it is necessary to devise a suitable system for storage and retrieval. Many companies are now using computer database systems for procedure management in order to reduce duplication and improve access to procedure data [175].

Validity of parameters and calibration of equipment. The validity of the procedure will depend on the possibility of establishing the same operating conditions when the process is used with the stated parameters, by a different operator in a different workplace using equipment and consumables of the specified type. If the parameters specified are ambiguous or ill-defined the resultant weld may well be inferior to those prepared at the procedure development stage.

For example in GMA welding when voltage is used as a control parameter the voltage reported should be the *arc* voltage measured as near to the arc as possible. Open circuit (or no-load) values may be easier to measure but are meaningless unless the static characteristics of the power source are also stated. In addition, since there may be a significant voltage drop in the welding cable at higher currents the *terminal voltage* of the power source, as indicated by the power source meters, is likely to be higher than the arc voltage and subject to variation with cable length and diameter. In the case of welding consumables, these should be specified according to a recognized national or international code which defines their composition limits and performance rather than by trade names. (In some cases the user may need to add specific limitations on performance and composition.)

The method of measurement of electrical parameters should be specified (e.g. root mean squared (RMS) or mean) and in some cases it may be necessary to specify the type of instrument and the measuring technique [176] (see also section 10.3 below).

The tolerances on procedure variables should be specified with due regard to the equipment capabilities and limitations. For example, new equipment will be calibrated in accordance with a manufacturing standard (e.g. BS 638 [177]) but the tolerances allowed are often quite wide; in fact for most equipment values of ± 10% of those indicated are permissible, and repeatability of settings between equipment cannot be guaranteed. Calibration of existing equipment may also be difficult, for example in the case of simple MMA equipment which often has poorly defined markings on controls which are subject to wear [178]. In GMAW equipment meters where fitted often become damaged and deteriorate in the normal welding environment and cannot be relied on for calibration purposes.

A further source of calibration error is the variation in output of conventional welding equipment with mains input swings which may be up to ± 10%; however, it is common for more modern electronic welding power sources to incorporate output stabilization which offers greater protection against uncontrollable input voltage fluctuations. In addition it is usually possible to obtain improved accuracy and repeatability from these electronic power sources as discussed in Chapter 3.

In view of the problems listed above it is necessary to check equipment which is intended for use on critical welding procedures with some external calibration device which itself has a known accuracy. The appropriate level of calibration will be determined by the application and a two-tier system has been proposed [179] and this is the subject of proposed codes of practice. This system consists of the following grades:

(i) *Grade 1*. This is the standard grade of calibration accuracy and as required by the power source design standards (e.g. BS 638).

(ii) *Grade 2*. This higher or 'precision' grade is intended for applications requiring greater precision, such as nuclear industry joints, mechanized, orbital and robotic welding systems.

Target requirements for each of these grades are shown in table 10.3.

It is not only the welding equipment which requires calibration; in the case of MMA welding where electrode temperature control may be very important, it is necessary to calibrate the electrode storage ovens and when preheat and post-heat treatment are involved, the heating equipment and the devices used for checking temperature need to be considered. In mechanized welding processes the travel speed and positional accuracy of the system will need to be checked regularly.

Table 10.3   Calibration requirements.

Parameter	Grade 1	Grade 2
Current	± 10%	± 2.5%
Voltage	± 10%	± 5%
Slope up/down time		± 5%
Pulse time		± 5%
Measuring instruments	± 2.5%	± 1%
Wire feed speed		± 2.5%
Calibration frequency	1 yr max.	0.5 yr max.

Monitoring of procedures. Once accepted it is essential that production welding operations are monitored to ensure that the procedure is being followed and that the required results are being achieved. Traditionally, testing of the completed fabrication using non-destructive test techniques (e.g. x-ray, ultrasonic, MPI and dye penetrant) have been used to check the weld quality and, if necessary, defects have been removed and the joint repaired. Progressive monitoring at an early stage of production can, however, prevent costly rework after final inspection. Routine monitoring should at least include checking the critical procedure variables (e.g. current, voltage, wire feed speed, temperature, consumable treatment, travel speed). Improved control may be achieved by use of the monitoring instrumentation described in more detail below.

The use of portable monitoring devices which comply with traceable standards of calibration is an effective way of ensuring that the optimum parameters are being used; it is particularly beneficial to use these same devices at the procedure qualification stage. In practice accurate high-quality analogue meters could be used for this purpose but these tend to be less robust than digital meters; however, digital meters must be used with care especially if the parameter being measured is not either a constant DC or pure sinusoidal AC waveform. Many recent monitoring and calibration devices incorporate a computer data logger which provides permanent records of the welding parameters and also allows the data to be transferred to a computer for permanent storage. (These systems will be described more fully in section 10.3.1(b).)

10.2.2 Summary—welding procedure control

Control of the welding may be an informal process; where the welding engineer or the welder is left to assess the requirements and select appropriate welding parameters, or a more formal approach may be adopted, in which the suitability of the welding procedure is tested and documented.

The traditional method of control by means of formal welding procedures depends on:

(i) establishing and proving satisfactory welding parameters by procedure trials and testing;
(ii) maintaining the same parameters in production;
(iii) monitoring by means of final inspection and NDT;
(iv) correction of errors by repair and rework.

It is assumed that if all the process inputs remain fixed a satisfactory repeatable output, in terms of weld quality, will be obtained.

Any input errors or *disturbance* in the process which cause deterioration of quality may not be noticed until final inspection. This may be considered to be an *open-loop* system since the quality of the output is not used directly to control the process; the control loop is closed by manual intervention to correct any errors but this is often carried out after the weld has been completed and the only means of correction is repair.

Full qualification and approval of procedures in this way is a costly and time-consuming process and is usually only justified when specific joint quality requirements must be achieved. However, in many cases it is essential to use these techniques to achieve an adequate measure of control, and satisfy quality assurance requirements. †

10.3   Monitoring

Measurement of welding parameters and calibration of equipment is essential when any method of control is to be applied. The requirements for formal welding procedure control have been discussed above but the availability of suitable monitoring methods is also a prerequisite of any automatic control system. Considerable improvements in the methods of monitoring welding processes have recently been made and some of the techniques available will be discussed below.

10.3.1   Welding parameters and measuring techniques

The techniques which will be discussed are:

(i) conventional meters;
(ii) computer-based instrumentation;
(iii) measurement of welding parameters;
(iv) stability measurement;
(v) dynamic resistance measurement;
(vi) deviation monitors;
(vii) vision systems.

---

† These techniques may need to be implemented to satisfy quality systems standards such as the BS 5700, ISO 9000 and EN 29000 series.

(a) *Conventional meters*

Analogue and digital measurement. Both analogue and digital techniques are applied in the measuring of welding parameters.

In analogue systems the signal to be measured is converted into an indication which changes continuously in response to variations of the input signal. In analogue meters the incoming voltage may be converted into the deflection of an indicating needle electromagnetically. Alternatively, for rapidly changing signals the indicator may be a beam of electrons which is scanned across a cathode ray tube screen by the deflecting coils of an oscilloscope.† Analogue measuring techniques give a good visual indication of the rate of change of parameters and are useful when assessing process stability. Analogue meters also give a good indication of mean output levels when the signal is subject to random variations (e.g. when measuring the current in MMA welding). Quantitative measurements may be made by reading a calibrated scale. The indication provided by this type of meter is continuous and the resolution depends on the scale and the care with which the readings are taken. In order to be accurate the meter scale needs to be large, but with conventional systems this implies a fairly heavy mechanical movement which may reduce the rate of response of the instrument. To obtain a permanent record of the indicated values it is necessary to record the movement of the indicator on paper or film; this may be achieved using a strip of continuously moving recording medium which is marked by a pen attached to a moving potentiometric carriage or a light beam which is deflected onto photosensitive paper by a galvanometer.

In digital measuring systems the analogue signal is converted to a number on a predefined scale before it is displayed. The resolution of the instrument will depend on the analogue-to-digital conversion (ADC) device used, an eight-bit device will provide a resolution of 256 increments whilst a 12-bit device will give 4096 and a 16-bit device will allow 65 536 increments to be resolved. If

---

† Oscilloscopes give a clear indication of the amplitude and frequency of regular, periodically varying signals but for transient or irregular signals single-shot instruments such as the computer-based systems described below must be used.

the full scale reading of the instrument is 256 V an eight-bit ADC would give a resolution of 1 V whilst a 16-bit device would provide 0.004 V resolution. For many welding applications eight-bit accuracy is adequate, particularly if the signal is conditioned to limit the full-scale value. The advantage of digital measuring systems is that they provide a direct numeric output and this may be stored or recorded electronically as described below. Digital systems are generally more robust than analogue meters but they are sometimes susceptible to electrical interference. The disadvantage of the digital approach is that it is difficult to interpret the digital display if the parameter being measured is fluctuating rapidly and the 'mean' values may be arrived at in several different ways (e.g. by electronic processing of the incoming signal or by calculation). This may lead to slight discrepancies between the values measured with different digital meters and may be responsible for large variations when compared with traditional analogue meters.

(b) *Computer-based loggers*
Computer-based data loggers as used in general process control and biomedical applications were originally only used in welding research but purpose-built monitors for calibration and control of welding processes are now available [180].

The principle of computer-based instrumentation is illustrated in figure 10.4. The analogue signal to be measured is amplified or attenuated by a *signal conditioning* circuit which consists of standard electronic components. The output of this stage may be electrically isolated from the remainder of the instrument by isolation amplifiers, and hardware filters may be incorporated to reduce electrical noise. Isolation is particularly important when high-voltage welding signals are being measured, where common-mode problems occur† and to avoid spurious signals when low currents or voltages are being monitored (e.g. from thermocouples).

† Common mode problems: when measuring welding current and voltage simultaneously it is possible to connect the instrumentation in such a way as to short circuit the output of the power supply. A high-current path may accidentally occur either in the interconnecting leads or, more seriously and less obviously, through the ground or earth connection of the instrument. Such faults can result in serious damage to welding instrumentation.

The analogue signal is digitized by an ADC similar to that used in digital voltmeters. The use of eight-bit ADC converters is again adequate for many welding applications providing the input is scaled to an appropriate level. In order to provide facilities for monitoring several welding parameters the conditioned analogue signal from a number of inputs may be scanned by a multiplexer before being passed to the ADC.

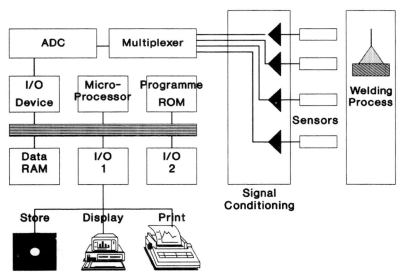

Figure 10.4   Principle of computer-based instrumentation.

The digital output from the ADC is sampled at a rate determined by the *clock* rate of the microprocessor and the control program. A range of programs and operating parameters may be stored in the program memory (usually an EPROM) and the appropriate sampling conditions may be chosen for a specific application. The microprocessor also determines where the digitized values are stored, how they are processed and whether they are displayed on some in-built indicator or transferred to an external device for storage or display. Systems of this type are capable of reading the instantaneous value of the input level every 100 μs or even faster. It is possible to store these instantaneous values in the RAM and when they are displayed they will reproduce the waveform of the incoming signal in much the same way as an oscilloscope, the difference being that the values are stored and may be examined repeatedly. The data may also be transferred to non-volatile memory (e.g. floppy disc, battery-backed RAM, tape or bubble memory) and hard copy of the waveform may be obtained with the aid of a printer or plotter.

In order to avoid inaccurate representation of the waveform, known as *aliasing*, the sampling frequency should be at least ten times that of the waveform being measured—with intersample times of 100 μs the sampling frequency is 10 kHz and signal frequencies lower than 1 kHz are accurately represented; this response is perfectly adequate for general assessment of the common welding waveforms found in pulsed GTAW and GMAW although higher sampling speeds are sometimes required when investigating high-speed transient events.

The welding waveform in memory may be displayed on a monitor and analysed. Some instruments also allow calculations to be performed on the waveform and discrete values of pulse parameters may be displayed. The number of data which may be captured is, however, restricted by the memory available and at high sample rates the random access memory will be filled very quickly. If detailed information concerning the waveform is not required the incoming signal may be processed in real time and more compact data file maintained. Two common techniques which have been used in welding applications are event monitoring and derived data storage.

In event-monitoring techniques only the data relating to those transient features of the waveform which satisfy certain criteria are recorded. A common application is the recording of pulse or short-circuit current peaks in GMAW. A threshold current is preset and only excursions of the current which exceed this threshold are recorded. Each time the current rises above the threshold the time, amplitude and duration of the event are stored. Since only events of interest are captured and the data concerning these phenomena are compressed, a large amount of relevant information may be obtained over an extended sampling period. A device capable of this type of data collection is shown in figure 10.5 and the application of this type of system for stability analysis is described in section 3.1.2.

The information may also be compressed by performing calculations on the raw data during the measuring process and storing only the results. Derived data such as mean current and voltage may be obtained for example for every thousand readings, the raw data may then be discarded and the averages stored in RAM. Alternatively, secondary process parameters such as heat input, or dynamic resistance may be calculated on-line in this way.

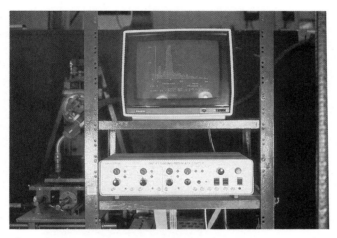

Figure 10.5   Typical general-purpose computer data logging interface.

A purpose-designed welding data logger/monitor which embodies many of the features described above is shown in figure 10.6. This instrument monitors arc voltage, current and wire feed speed and has facilities for waveform capture and analysis and the capability of monitoring mean values for each weld run. In the case of waveform capture variable sample rates and pre-triggering facilities are available. The output may be directed to:

(i) an external oscilloscope for general analysis and measurement of the waveform;
(ii) an internal display which gives a digital read-out of the values of current or voltage identified by a movable cursor on the oscilloscope screen;

(iii) an internal printer which prints the calculated values of peak and background current and voltage, peak and background time and mean current and voltage;

(iv) a removable battery-backed RAM cartridge which may be used to store data and transfer it to a personal computer for more detailed display and analysis. Using this facility the measured data may be presented as a welding procedure, a weld costing or a comparison of the measured values with a preset procedure.

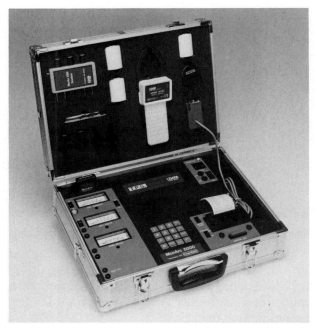

Figure 10.6   Purpose-designed welding data logger for steady and transient monitoring of arc processes. (Courtesy Data Harvest.)

Portable instruments of this type provide a useful means of calibration, monitoring and problem solving in welding applications. These systems provide information after the weld being monitored is complete but the techniques employed may also be applied to permanent data recording and control in real time as described below (under *'Speed and time'*).

The recent development of low-cost remote computer-based data loggers has made it possible to record welding parameters in the welding environment, store the data in battery-backed memory and interrogate the results by downloading the data through a serial link either via a long fibre optic communication cable or by direct connection of the remote data collection unit to a suitable computer (figure 10.7). This system has been used for measurement of heat-affected-zone temperature cycles, thermal testing of welding torches and monitoring of current and voltage [181]. The improved availability of high-speed analogue-to-digital cards for personal computers has also made it possible to develop cost-effective purpose-designed welding data loggers suitable for research, production and quality monitoring applications [182].

Figure 10.7 Portable   computer   and   battery-operated compact data logger.

(c) *Normal welding parameter monitoring*

Temperature. The high temperatures experienced in welding may be measured using temperature-indicating paints or crayons, fusible indicators, bi-metal thermometers, thermocouples, infrared thermometers or by thermal imaging.

Temperature-indicating paints and crayons change colour in response to the surface temperature of the material. They are convenient for site use but their accuracy depends to a certain extent on the care with which they are applied and the interpretation of the colours.

Fusible indicators melt and in some cases change shape when they reach a prescribed temperature. They give a clear indication that a certain temperature has been reached but little guidance on the actual temperature of the material during the heating and cooling cycle.

Bi-metal thermometers are contact devices which rely on the response of a bi-metal strip to temperature, deflection of the strip being converted mechanically into movement of a needle. The indication of temperature is continuous and easily read but the accuracy is often less than 10% of full scale.

A more accurate means of temperature measurement uses a thermocouple as a sensor. The thermocouple junction produces a small voltage which is proportional to its temperature; this voltage may be measured directly by an analogue meter or it may be digitized and displayed directly or stored as a permanent record.† The signal from the thermocouple is usually small and may need to be amplified; in addition corrections need to be applied for the cold-junction temperature and the current must be converted into a temperature reading. These functions are most easily performed electronically and since most temperature signals change relatively slowly digital meters often provide a convenient output display.

Non-contact measurements may be made by detecting the infra-red radiation from the material and thermal imaging cameras which give an indication of the temperature profile over the area of interest are available. These devices are, however, relatively costly and their use is at present restricted to research and automatic sensing systems.

Current. Instantaneous current values may be displayed on an oscilloscope and the overall shape of both AC and DC waveforms

† Thermistors and resistance thermometers may be used to produce similar electrical output signals but thermocouples are usually more suitable for the range of temperatures normally encountered in welding and ancillary operations.

may be examined. However, the input of the oscilloscope amplifier will usually be calibrated in $V\,cm^{-1}$ and the maximum level of input will often be around $5\,V\,cm^{-1}$. In order to convert the normal high currents used in welding into a low-voltage signal one of the following devices is normally used: current shunt; current transformer or Hall-effect probe.

The *current shunt* is a high-power, low-value resistance which is placed in the circuit through which the current to be measured is flowing. It produces a low-voltage signal (e.g. 50–200 mV) proportional to the current passing through it. Whilst this is suitable for measuring relatively slowly changing waveforms most shunts do have some inherent inductance which limits the rate of change of the signal and distorts the observed waveform. Resistive heating of the shunt can also lead to inaccuracies. For improved response and accuracy water cooled *non-inductive* shunts have been devised but these are usually costly and less convenient to use.

For AC waveforms *current transformers* may be used to reduce the output to a suitable level. These usually take the form of a toroid or coil which is placed around the conductor which is carrying the current to be measured, although low-cost devices which employ a clamp-type construction are available. Again these devices may produce some distortion of rapidly varying waveforms.

Hall-effect probes are based on semiconductor elements which respond to the magnetic field which is produced when current passes through a conductor. They are capable of detecting and indicating DC and AC current and have excellent frequency response which enables them to be used to detect transient phenomena in rapidly changing waveforms.

The use of oscilloscopes is, however, generally restricted to research and the servicing of welding equipment. For production purposes a quantitative, single-value measure of current is normally required. Steady DC current may be measured with analogue, moving-coil and digital meters. For DC currents the value indicated is the steady DC value or in the case of fluctuating current waveforms the mean value. Analogue moving-iron meters indicate the RMS value of a fluctuating signal. For the waveform shown in figure 10.8 the mean value is given by

$$I_m = (I_1 + I_2 + \cdots + I_n)/n \qquad (10.1)$$

where I_1 to I_n are the currents at regular intervals and n is the number of values measured. Alternatively the average is given by the area under the curve divided by the equivalent length of time measured along the x axis. For a rectangular waveform this is given by

$$I_m = \frac{I_p t_p + I_b t_b}{t_p + t_b} \qquad (10.2)$$

where I_p is the peak current, I_b is the background current, t_p is the peak time and t_b is the background time.

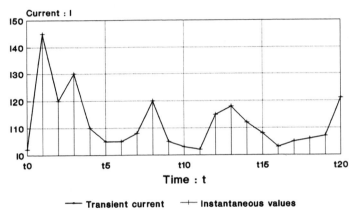

Figure 10.8   Mean of fluctuating current waveform.

The mean value of a balanced AC waveform would, however, be zero but its effect may be measured by considering the steady DC value of current which would give an equivalent amount of resistive heating. This value is known as the RMS current and is given by

$$I_{RMS} = \frac{\sqrt{(I_1^2 + I_2^2 + \cdots + I_n^2)}}{n}. \qquad (10.3)$$

Alternatively the alternating current may be rectified and the mean value of the resultant DC waveform may be taken, this value is referred to as the *mean absolute* value (MA). For steady DC and symmetrical square waveforms the mean absolute value of current is equivalent to the RMS value but for any other waveform these values will differ. For example with a pure sinusoidal waveform the RMS value is 1.111 × the mean absolute. † In the case of rectangular waveforms there may be a substantial difference in MA and RMS values as shown in table 10.4. It is most important to specify the methods which are being used to measure current and to adopt the same techniques when comparing equipment or processes.

Table 10.4 Difference between mean and RMS current values for rectangular waveforms.

Pulse current (amps)	Background current (amps)	On/off time ratio	Mean absolute value (amps)	True RMS value (amps)	Scaled RMS value* (amps)
600	0	1/1	300	424	333
550	50	1/1	300	390	333
500	100	1/1	300	360	333
450	150	1/1	300	335	333
300	300	1/1	300	300	333
600	0	1/3	150	300	167

Note: 1). The scaled RMS value (*) is typical of the reading obtained with a low cost meter which derives the RMS from the measured mean absolute current.

Adapted from; Robinson.P,Rogers.P.F. Measurement of welding current, letter to the editor, Joining and Materials,2(7), July 1989, p314.

Voltage. The voltages used in welding usually need to be attenuated before measurement either by oscilloscope or by analogue or digital meters. Most metering systems do, however, incorporate suitable attenuation for the normal levels of voltage to be measured. The system of measurement and type of instrumentation must again be specified since variations will occur between different waveforms or instrumentation as described above. The simplest method of checking the output of power sources is by the use of an ammeter and voltmeter or a combined instrument as shown in figure 10.9. This may comprise a Hall-effect current sensing device and a digital voltmeter.

† The RMS value divided by the mean value of the current is known as the *form factor* of the waveform, i.e. 1.111 is the form factor of a pure sine wave. The peak current divided by the RMS value gives a factor known as the *peak or crest factor*, the higher the crest factor the greater will be the difference between mean absolute and RMS values.

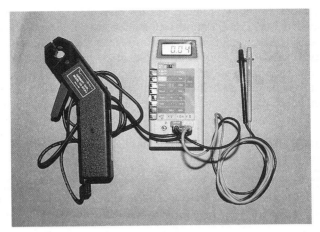

Figure 10.9 Digital voltmeter and Hall-effect current probe, suitable for arc current and voltage measurement.

Speed and time. Linear travel speed and wire feed speeds are often measured manually by timing over a measured travel distance or amount of wire fed. Measurement using electrical and mechanical tachometers may also be used but it is necessary to translate the linear motion to rotational movement. Suitable sensors are available as shown in figure 10.10; these usually use

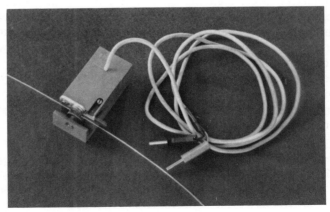

Figure 10.10 Wire feed speed sensor (tacho generator).

either slotted disc optical tachometers or small DC generator devices. The voltage generated may be displayed on analogue or digital instruments or in the case of the slotted disc encoder the frequency pulses may be easily converted into rotational speed.

(d) *Stability monitors*

The measurement of process stability is of interest when monitoring the performance of the consumable electrode arc welding processes. It may be used to assess the process performance during production as well as assisting the development of consumables and equipment.

The operating performance of conventional SMAW and GMAW processes may be evaluated by welding trials conducted by an experienced welder. These subjective assessments require considerable skill, are time consuming and the results are often inconsistent. Objective measurements of stability may be achieved by examining the statistical variation of various operating parameters. A wide range of stability criteria have been developed [183–194] and some of these are listed in table 10.5.

In dip transfer GMAW the standard deviation of the arc time, the standard deviation of the peak current and the ratio of arc time to short-circuit time give a good indication of stability and are easily measured using computer-based instrumentation. Figure 10.11 shows histograms of arc times (measured with the event-monitoring data logger shown in figure 10.5) for two nominally similar GMAW consumables. The corresponding bead profiles are shown in figure 10.12. It is apparent that the consumable with the better stability, as indicated by the lowest standard deviation of arc time, gives the best operating performance, lowest spatter levels and better bead shape.

Although this type of monitoring has been used successfully in the research environment few commercial stability meters have been produced. It must also be remembered that arc stability, though important, may not be the main criterion for weld quality assessment; in GMAW welding in particular optimum stability may be sacrificed if an improvement in fusion characteristics can be achieved. The possibility of using this type of measurement to detect unacceptable deviations in operating performance is, however, an interesting prospect and may be exploited in on-line quality monitoring systems as discussed below.

Table 10.5   Stability criteria.

Criterion	Application	Reference
$C_{crit} = S^2/X_m^2$	CO_2-shielded GMAW	[183]
As above but deviation of current peaks used.	AC MMAW	[184]
$r = w/W$ and $= i/I$	DC power source evaluation	[185]
$K = \dfrac{\text{mean cycle duration}}{\text{deviation cycle duration}}$	Element recovery in GMAW	[186]
Standard deviation of current peak values	General process monitoring	[187]
K factor, as above	Shielding-gas evaluation for GMAW stainless steel	[188]
Standard deviation of arc time	Shielding-gas evaluation, GMAW plain carbon steel	[189]
Probability distribution of short-circuit and arc time	CO_2-shielded GMAW	[190]
Statistical analysis of arc and short-circuit times	GMAW spatter level investigations	[191]
$\dfrac{\text{standard deviation } t_{arc}}{\text{average arc time}}$	GMAW dip transfer	[192]
$W_s = W_a + W_r + W_p$ based on statistical analysis and correlation	GMAW	[193]
Statistical analysis of short-circuit cycle time	MMAW and GMAW	[194]

Key: S is the standard deviation of time between short circuits measured on a voltage trace. X_m is the mean value of the time between short circuits. w is the minimum value of power in the arc, just after droplet transfer. W is the average power under steady conditions. i is the minimum value of current and I is the average value of current under steady conditions. W_a is a stability criterion based on factored values of the standard deviation of arc time, short-circuit time, arc current and peak current. W_r is a stability criterion based on the ratio of the observed resistance during arcing and the average resistance. W_p is a criterion based on the ratio of observed power during arcing over the average power.

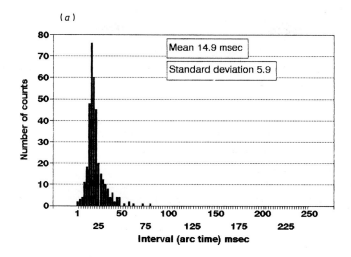

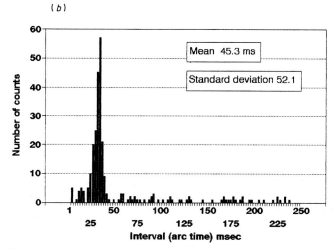

Figure 10.11 Histogram of arc time for dip transfer GMAW with (*a*) a good wire and (*b*) a poor wire.

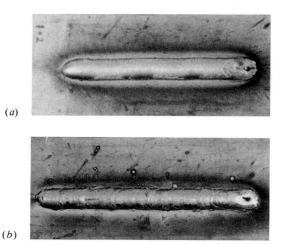

(a)

(b)

Figure 10.12   Weld bead appearance for tests shown in figure 10.11. Good wire (a) gives smoother finish and low spatter level in comparison with poor wire (b).

(e)   Deviation monitors

Dynamic resistance. The quality of resistance spot welds may vary due to the surface conditions of the plate being welded, electrode wear, current shunting and transient changes in the welding parameters. Various methods of monitoring spot weld quality in order to detect the effect of these variations have been proposed but the dynamic resistance technique is probably the most suitable for industrial application [195,196]. In this system the instantaneous welding current and voltage are measured and the resistance of the metal between the electrodes is calculated. If computer-based techniques are used to collect the data the resultant dynamic resistance curve may be plotted on the monitor screen immediately after the weld has been completed. The normal dynamic resistance curve for a satisfactory weld in plain carbon steel is shown in figure 10.13. The curve may be divided into three zones:

(i) Zone A: the area corresponding to the reduction of contact resistance.

(ii) **Zone B:** the area corresponding to resistive heating and increasing resistivity of the material.

(iii) **Zone C:** the area corresponding to a reduction in thickness and weld nugget growth with consequent steady reduction in resistance.

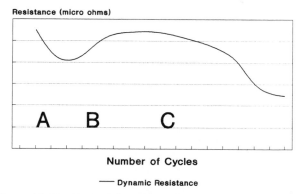

Resistance (micro ohms)

A   B   C

Number of Cycles

—— Dynamic Resistance

Figure 10.13 Typical dynamic resistance curve for a normal spot weld.

Changes in shape of the curve indicate deviations from the normal weld nugget formation and in the limit a defective weld. The computer-based dynamic resistance instrument [197] shown in figure 10.14 may be used to monitor the performance of the process in the following manner.

The welding parameters are set and a test specimen is welded with the monitor set in its 'test' mode. The welded specimen is checked visually and by destructive testing and if a satisfactory weld has been achieved the dynamic resistance curve which has been obtained is stored as a *master* for the application (if the weld is unsatisfactory the parameters are adjusted and a new test is conducted). With the equipment switched to its '*monitoring*' mode the dynamic resistance curves of subsequent welds are compared with the master curve and deviations which exceed preset limits are reported, used to initiate an alarm or even used to initiate the printing of a '*reject*' label. Welding current, voltage and time are also checked against preset values to ensure that satisfactory tolerances are maintained.

The record welding parameters may be stored in the monitor's memory and measured data may be stored on a non-volatile memory device (e.g. a floppy disc). This type of equipment may also be configured as a series of local monitors mounted on individual welding machines which feed the collected data back to a central computer for permanent storage or analysis. Distributed computer-based dynamic resistance monitors of this type are suitable for on-line surveillance of automation systems and robotic welding stations [198].

Figure 10.14 Weldskan dynamic resistance monitor for resistance spot welding. (Courtesy Techno Instruments Ltd.)

Arc welding deviation monitors. Arc welding processes may be monitored using similar techniques to those described in section 10.3(d); however, in this case it is usually necessary to monitor a continuous weld rather than a discrete spot. A computer-based device suitable for quality assurance of automated GMAW welds is shown in figure 10.15 [199]. Arc current and arc voltage signals are monitored and analogue circuitry is used to derive short-circuit resistance and a radio-frequency (RF) component of the voltage waveform. The data are analysed by comparing the sample data with preset limits for voltage, current, RF component and dip

Figure 10.15   Arc Guard monitoring system (Cranfield).

resistance. To allow continuous updating of the data collected the four input channels are sampled at approximately 15 kHz and the data are analysed on-line. In normal dip transfer GMAW welding operations an assessment equivalent to each 2 mm increment of weld length is stored, passed to an external computer or used to initiate weld quality signals. Using real-time analysis and a knowledge of the interrelationships between the monitored variables and potential weld failure modes [200] the unit is able to predict a large number of weld quality deviations from the limited input data. For example, the dip resistance may be used to indicate variations in torch-to-workpiece distance, whilst excessive voltage in conjunction with a satisfactory dip resistance may indicate wire feeding problems. The RF component of the voltage waveform is used to indicate process stability and in particular the disturbance of gas shielding. Some of the common quality deviations which can be monitored with a device of this type are indicated in table 10.6.

Table 10.6   Quality problems identified by 'Arc Guard'.

Current high or low
Voltage high or low
Shielding-gas failure
Poor arc stability
Wire feed slip
Wire run-out
Stick-out long/stick-out short
Midweld power loss
Welding supply failure

When applied to an automated welding system a *pass/fail* assessment may be made for each increment of weld and as a percentage of the individual weld length, in addition the results of the assessment may be used to initiate a controlled shutdown of the welding cell or operations such as torch cleaning.

These monitors are usually provided with a serial output port (e.g. RS232 or 422) to enable data to be downloaded to a central computer from a number of welding cells where they may be assessed and compared and production statistics and overall performance may be assessed.

Monitoring systems which provide on-line indication of deviation from some preset parameter are sometimes called *programmable error monitors* [201] and these may vary in sophistication from microprocessor-controlled devices to simple low-cost electronic alarms [202].

Statistical process control. The ability to transfer or download large volumes of data to a personal computer allows statistical process control techniques to be used [203,204]. Statistical process control is defined [205] as:

> The monitoring and analysis of process conditions using statistical techniques to accurately determine process performance and preventative or corrective actions required.

It has been shown [206] that the laws of probability may be used to define predictable limits of variation in quality characteristics, and excursions of the observed data outside these limits should be taken as an indication of potential quality problems. In any welding process there will be a natural scatter in the observed parameters resulting from inherent process characteristics and environmental conditions. This natural random scatter is known as 'common cause' variation. Under many circumstances if the process is operating normally and sufficient data are collected the values will fall within a standard probability distribution similar to that shown in figure 10.16. If the observed parameters fall outside the established distribution then abnormal or 'special cause' process disturbances are responsible. By monitoring the process during normal operation, calculating the sample mean (X_m), the range of each sample (R) and the standard deviation (σ) it is possible to set upper and lower control limits (UCL and LCL, usually = 3 σ). If any of the parameters subsequently measured fall outside these control limits it may be regarded as an indication of abnormal performance. These techniques may be applied to any welding process and have been implemented for both the resistance and the arc monitoring systems described above. Figure 10.17 shows the screen of a resistance spot welding monitor. In the upper section is the sample of the normal population (each point represents a weld) and at the left of the screen the corresponding distribution which has been derived from the data. In the lower half

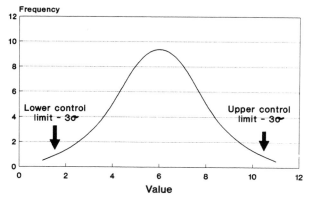

Figure 10.16 Normal probability distribution.

of the screen the results of monitoring real welds and the adverse trend towards the upper control limit is shown.

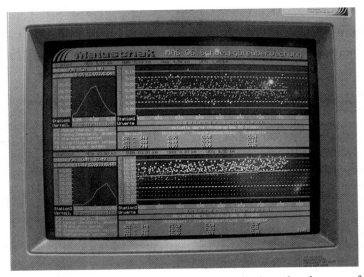

Figure 10.17   Dynamic resistance monitor display, showing use of the statistical process control technique. (Courtesy Matuschek Industrie-Elektronik, Germany.)

With the arc monitor, control limits may be established in a similar way by collecting data from normal welding operations and a control chart may be produced. The effect of abnormal conditions may then be identified by plotting subsequent data on the chart. The charts shown in figure 10.18 show the effect of deliberate process disturbances. In the case of gradual reduction of shielding gas flow the % POOR values† indicate an adverse trend which exceeds the control limit at subgroup 5. At this stage the gas flow had been reduced from its original value of $12\,l\,min^{-1}$ to around $7-8\,l\,min^{-1}$ and although the effect was just discernible by

† % POOR is derived by expressing the number of weld segments which have failed to remain within the preset limits over the total number of segments in a weld as a percentage.

visual inspection of the finished welds the welds were still acceptable for the intended application. By the time the gas flow had
been reduced to $2\,l\,min^{-1}$ (subgroup 9) the welds were 100%
POOR and visually unacceptable. The charts do provide early
warning of unacceptable performance and alert the user to adverse
trends before the weld quality reaches an unacceptable level.

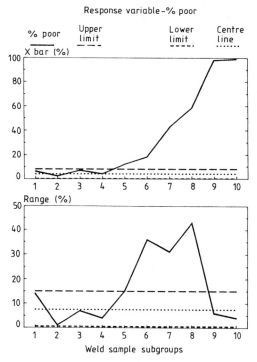

Figure 10.18 Control chart output from Arc Guard (figure
10.15) showing indication of poor gas shielding.

Visible spectrum monitoring. The use of a spectrographic
technique for real-time monitoring and control of weld quality has
been reported [207]. The system can detect changes in the chemical
composition of the arc and in particular the loss of shielding gas,
the increase in hydrogen in the arc atmosphere and changes in the

flux composition during the FCAW process. Although the equipment used in the feasibility study was costly and complex it is
envisaged that once the specific monitoring requirements are identified simple band-pass filters could be used to produce a compact
welding monitor.

(f)   *Process monitoring—summary*

Monitoring techniques are available for calibration and trouble-
shooting in welding processes; these vary from simple meters to
on-line computer-based systems which collect and report the
status, trends and production statistics on a large number of
welding cells.

Regardless of the type of monitoring system used it is important
to ensure that the correct measuring techniques are applied and the
procedure and type of instrumentation are recorded.

Computer-based monitors provide facilities for on-line quality
assurance which can make a significant contribution to the cost
and reliability of the welding operation and reduce the need for
post-weld testing [208,209]. It is reported, for example, that one
manufacturer has used on-line data logging to reduce post-weld
destructive testing by 50% and saved over \$90 000 per year [210].
The data obtained may also be used for optimizing the welding
process and preventative maintenance programmes. Significantly,
many quality systems standards recognize the use of continuous
monitoring as an important control technique; BS 5750 [211], for
example, states:

> 'Continuous monitoring and/or compliance with the documented
> procedures' are required to ensure that the specified requirements
> are met.

It is important to note that although these systems may be
relatively sophisticated and provide clear indications of process
deviations they require manual intervention to correct the process
performance.

10.4   Automatic Control Techniques

Using suitable monitoring systems and mechanized welding it is
possible to correct process deviations automatically without the

need for manual intervention. All of the methods involved rely on the application of a control systems approach and it is important to understand the basic principles of control as applied to industrial processes before discussing the common applications.

10.4.1 Control systems

The fabrication system may be illustrated as shown in figure 10.19. The inputs to the system are materials and consumables which are converted by the process into finished welds. Control systems of this type are divided into two types [212]:

(i) Open-loop, in which the output has no direct effect on system control as shown in figure 10.20(a);

(ii) Closed-loop, in which some signal derived from the output is used to directly control the system as shown in figure 10.20(b).

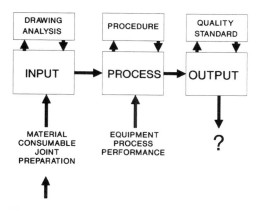

Figure 10.19 Normal control techniques used in welded fabrication.

The welding of joints to a predetermined procedure is effectively an open-loop system. Although every effort may be taken to ensure that the predetermined parameters are maintained (e.g. by equipment calibration and monitoring), any corrective action taken as a result of post-weld inspection is indirect and occurs some time after the process is complete.

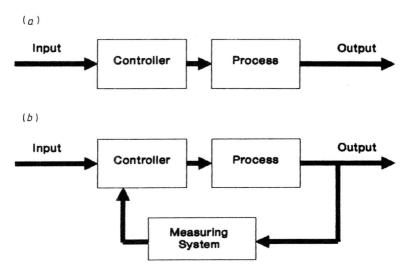

Figure 10.20 (*a*) Open-loop control system. (*b*) Closed-loop control system.

Closed-loop systems are also referred to as *feedback systems* since a signal derived from the output is fed back for comparison with a reference parameter. The result of this comparison is an error signal which is fed to the controller in such a way as to reduce the error. The error correction system or controller may be a manual operator. In dip transfer GMAW operations, for example, the welder often *tunes* the process performance by adjusting the wire feed speed, voltage or inductance in response to the sound of the arc; in this case the sound is an indication of the short-circuit frequency. In more advanced control systems the human operator is replaced by an automatic control which monitors the output and corrects the system continuously using an error signal derived by comparing a suitable output parameter with the preset reference. In automatic closed-loop control systems the rate at which the error is corrected is critical, as slow response rates or overcorrection may both lead to instability and the design of suitable control strategies will have a significant effect on system performance.

Several variations of closed-loop control are used in welding applications. These may be referred to as: direct control; indirect control; adaptive control; and learning control.

Direct control implies the measurement and feedback of the parameter which it is sought to regulate. For example in the case of a power source the current may be regulated by monitoring its output level as described in Chapter 3. Unfortunately in many welding situations it is not possible to use direct control and *indirect* systems based on the measurement of secondary variables must be used. Several indirect control systems will be described below but typical examples are; weld bead temperature for penetration control and dynamic resistance for resistance spot weld geometry.

The term *adaptive control* is often used to describe recent advances in welding process control but strictly this only applies to systems which are able to cope with dynamic changes in system performance [213]. Adaptive systems should have the ability to self-adjust in accordance with unpredictable changes in operating conditions. It has been pointed out [214] that in welding applications the term adaptive control may not imply the conventional control theory definition but may be used in the more descriptive sense to explain the need for the process to adapt to the changing welding conditions.

Learning control systems are systems which progressively improve their performance by experience—the most common example is the human operator—but computer-based learning systems are being developed for welding applications.

10.4.2 Basic requirements for closed-loop control

For successful closed-loop control in welding, appropriate feedback signals must be obtained; these must be related in some way to the parameter to be regulated. A suitable error signal must be generated and the system must be able to respond to the required correction. The feedback signals may be obtained by monitoring the normal welding parameters or by employing sensors or transducers which supply indirect data. The relationship between the output and the control parameters may be defined by a mathematical model, algorithm or equation. The way in which the control parameters are corrected is defined by the control strategy and its effectiveness will largely depend on the response rate of the system, and the introduction of electronically regulated power sources has enabled major improvements to be made in this area.

The main applications of automatic process control in welding are line following/seam tracking, arc length control, penetration control and on-line quality control. These are discussed below.

(a) *Line following*
Many line following and seam tracking systems have been developed for welding and these may be classified according to the sensor systems used.

Tactile sensors. The simplest tactile sensor is the spring-loaded guide wheel which maintains a fixed relationship between the welding head and the joint being followed as shown in figure 10.21. This low-cost system may be used very effectively on simple mechanized tractors and gives reduced set-up times and improved consistency. A more sophisticated tactile sensor utilizes a probe which converts displacement to an electrical signal which is fed back to the motorized welding head causing the torch to be repositioned to follow the seam. An application of this system to the welding of Archimedean screw conveyors is shown in figure 10.22.

Whilst these systems are inherently simple, tracking errors may result if the surface condition of the plate is poor or in multirun situations where the probe must track a convex weld deposit. These problems may be minimized by using special probe tips (e.g. to cope with tack welds).

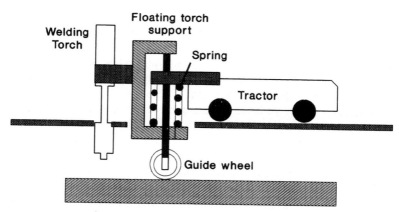

Figure 10.21 Tactile torch height control.

Figure 10.22 Tactile sensor following a complex seam—
GMAW fillet weld on a screw conveyor.

Pre-weld sensing/joint location. In fully automated and
robotic welding cells the seam may be located prior to welding
using tactile or non-contact sensors. A common technique is to
utilize the end of the GMAW electrode as a contact probe [215]. In
robot applications the torch moves to a taught point in the vicinity
of the joint (figure 10.23), the head is then moved from the taught
point along the z axis until it comes into contact with the workpiece, (contact is detected by a high-voltage, low-current DC signal). The contact point is confirmed by a short slow sweep along
the x axis away from the contact point and back. The torch then
travels along the x axis until it contacts the other plate and again
a short, slow-speed traverse away from and back to the second
contact point is performed. From the measured data the joint
intersection point is calculated and the torch is moved to an appropriate start point.

Similar systems have been used for applications involving portable robots in shipbuilding. In this case the robot uses a tactile
search routine to check its position within the welding cell and
adjusts the program datum point to correct any errors [216].

Pre-weld sensing with optical sensors has been applied to large
aluminium structures fabricated by GMAW [217] where due to the

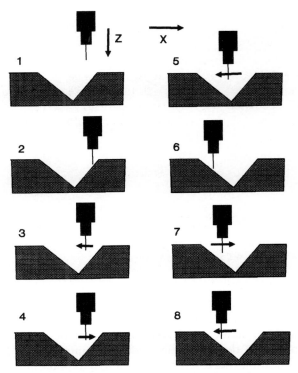

Figure 10.23   Tactile sensing—seam location.

complexity of the component a significant tolerance build up may occur. In this case a laser vision system (see '*Structured light/vision sensors*' below) was used to track and correct the weld path prior to welding.

Through-arc sensors. Through-arc sensing techniques utilize the change in one or more of the electrical parameters of the arc during oscillation of the torch tip to locate the joint position. These techniques are usually employed in fillet, heavy-section V butt and narrow-gap welding situations. If the welding head or the arc is oscillated laterally across the seam the arc length or electrode-to-workpiece distance will decrease as the joint edges are reached (figure 10.24). In GTAW with a constant-current power source the voltage will decrease at the outer edges of the weld, whereas when a constant-voltage power source is used for GMAW or SAW the current will increase. The limit of torch travel may be controlled by the change in parameters such that the centre of oscillation is always maintained on the joint axis. In GTAW and plasma welding the arc may be oscillated magnetically to provide the joint scanning required for through-arc seam tracking [218].

These through-arc tracking systems are now commonly available on welding robots and have also been applied to the orbital GTAW of pipelines [219].

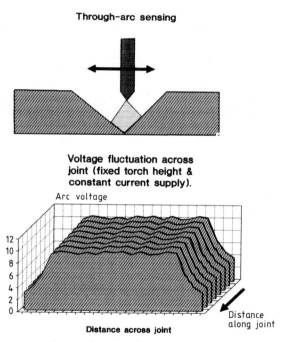

Through-arc sensing

Voltage fluctuation across joint (fixed torch height & constant current supply).

Arc voltage

Distance along joint

Distance across joint

Figure 10.24   Through-arc sensing.

Inductive sensors. Inductive proximity sensors may be used to track fillet welds as shown in figure 10.25, whilst eddy current sensors may be used to follow square butt joints. Inductive sensors are relatively compact, robust and unaffected by the visible radiation from the arc but spurious signals may be generated by magnetic fields in the workpiece or the holding fixture. They have been used successfully for GMAW seam tracking of butt, lap and fillet welds in robotic applications.

Figure 10.25   Inductive proximity sensors tracking a GMAW fillet weld.

Photoelectronic sensors. The simplest photoelectronic seam tracking sensors use a photoemitter and collector which are directed at a clearly defined joint line. In some cases it is necessary to employ a tape laid parallel to the seam as shown in figure 10.26 to ensure that a clear signal is obtained. These systems are relatively low cost but require a clearly defined joint line and may be affected by arc glare. A system which involves scanning the joint ahead of the welding torch with an infrared transmitter/ receiver assembly has been described [220] and this gives improved joint line resolution without the need for a pre-placed tape. An alternative system illuminates the joint area 30 mm in front of the torch with conventional lamps, infrared diodes or laser diodes; the reflected light is measured by a photodiode array or line transducer consisting of 256 diodes [221].

Structured light/vision sensors. In most structured light/vi-

Figure 10.26 Optical tracking system using pre-laid tape parallel with the weld seam.

sion systems a small CCD† video camera is used to capture the image of a line of structured light which is projected onto the weld seam in a transverse direction. The principle is illustrated in figure 10.27. The structured light may be obtained from a conventional lamp and projection lenses but more commonly a low-power helium:neon or diode laser is used. The laser stripe may be generated by optical techniques or the use of a mirror to oscillate the laser across the seam. The joint profile will be reproduced in the video image as shown and this information may be digitized and the difference between the real and a reference image may be used to generate an error signal which is used by the control system to correct the lateral position of the torch. To avoid interference from the arc a band-pass filter corresponding to the wavelength of the light source is used; in most cases this will be at the red or infrared part of the spectrum where arc radiation levels are low. This type of system has been used successfully in robotic and automatic applications for tracking lap, corner, fillet, V butt and multipass

---

† CCD cameras are preferred for these applications due to their compact size, robustness and the susceptibility of vidicon tubes to damage from intense light sources such as welding arcs.

butt welds. Applications include:

(i) Pressed steel components for automotive suspensions with complex three-dimensional seams. Welded with dip transfer GMAW. Weld placement being maintained to an accuracy of 0.5 mm.

(ii) Welding of 0.4 mm thick Inconel tubing by GTAW for aerospace applications (tracking accuracy of 0.1 mm).

(iii) Welding of multipass joints in thick-section steel using FCAW.

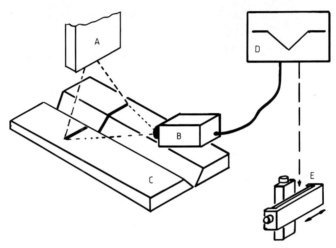

Figure 10.27   Principle of laser structured light seam tracking system: A, laser light source; B, CCD video camera; C, joint; D, image analysis; E, torch positioning system.

The original laser/vision systems were relatively bulky and costly but recent developments have allowed these tracking systems to be made more compact as well as improving the response rate. A typical system is shown in figure 10.28. The technique may also be used for pre-weld inspection and adaptive control of the welding variables [223] as discussed in section 10.4(d) '*Bead geometry prediction*'.

Direct vision sensors. The weld area may be viewed by means of a video camera without the use of structured light. With a single

Figure 10.28 'LaserVision' structured light guidance system for three-axis (x, y, z) control. (Courtesy of MVS Modular Vision Systems Inc., Canada.)

camera it is difficult to obtain three-dimensional information concerning the joint profile, but by using optical viewing systems concentric with the torch the relative position of the joint may be determined as shown in figure 10.29 [224]. By analysing the position of the joint line in a window of the video image the lateral error of the torch position may be determined. Although the torch assembly is rather complex this approach has the advantage that the arc radiation is effectively blocked by the electrode, and by measuring the weld width from the video image an additional feedback signal may be obtained for control of welding speed. The system was originally developed for GTAW; it has been reported [225] that the technique can in principle be used with GMAW.

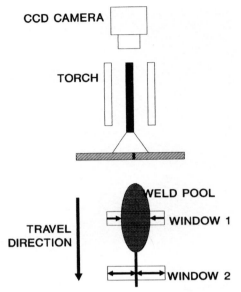

Figure 10.29 Concentric video monitoring system for GTAW.

Ultrasonic sensors. Ultrasonic sensors may be used to detect the unfused joint line in the parent plate. The principal problem involved in using these techniques has been the need for intimate contact between the ultrasound transducers and the plate, but a recent development [226] using pseudo-immersion probes allows the joint line to be detected to an accuracy of 0.2 mm, with normal plate surface conditions.

Chemical composition sensor. A novel approach to seam tracking is the use of a component of the arc spectrum to detect the presence of particular chemical species [227]. The arc emission is collected by a fibre optic and passed to an on-line analysis device which monitors the level of the signal produced by specific element. In a dissimilar metal joint this signal indicates the lateral position of the arc and may be compared with a reference level to generate an error signal for correction of the torch position. In joints of uniform composition tracer elements may be introduced in the form of fusible inserts.

(b) Arc length control

Control of arc length is important to ensure consistent heat input, constant melting rate, stable process performance and adequate shielding. Arc length may be maintained by good joint preparation, careful torch positioning or tactile sensors of the type described in section 10.4(a). Several closed-loop systems have also been used for arc length control.

Voltage measurement. In the GTAW process constant-current power sources are normally used and the arc voltage gives a clear indication of arc length at a given current. These systems directly measure arc voltage, compare the actual value with a reference and control the torch height to correct the error.† The arc voltage depends on current, shielding gas, workpiece composition, polarity and electrode vertex angle and the system reference must be set for the particular conditions being used. When the system is applied to pulsed GTAW or plasma welding the measurement of voltage must be synchronized with the pulsed waveform in order to avoid instability. Arc voltage sensing is probably the most common system of arc length control, no external sensors are required and the technique may be implemented using conventional analogue electronics or digital microprocessor control.

Sound measurement. Control of arc length in GTAW arcs may be achieved by applying an oscillation frequency to the arc and measuring the sound level. The sound pressure has been found to increase with arc length as shown in figure 10.30 [228] and this may be used to monitor changes in arc length and produce signals for arc length correction. One problem in the practical application of this technique is the interference from ambient noise; this may be minimized by operating at frequencies outside the normal audible range. Problems may also arise with high-frequency switching techniques used in modern power supplies but filtering may be used to overcome this.

Photoelectric sensors. Simple optoelectronic sensors comprising an infrared emitter and collector are used as proximity sensors

---

† Arc length control by means of voltage is often referred to as AVC or arc voltage control.

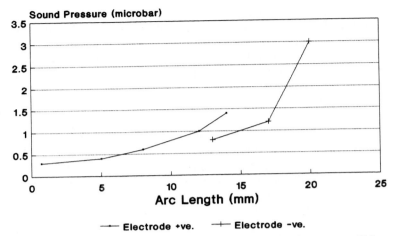

Figure 10.30 Effect of arc length on sound pressure for GTAW. (After Habil *et al* [228].)

for many non-welding applications. These devices may be used for torch height sensing as shown in figure 10.26. They are of low cost, relatively robust and unaffected by transient arc voltage instability.

Laser range finders. Laser range finders may be used as torch height sensors for robotic and automatic applications. These devices employ a low-power solid state laser and can measure distances very accurately using triangulation techniques. Due to their high cost they are rarely employed simply for height sensing and usually incorporate seam location and tracking facilities.

Light and spectral radiation sensors. The total arc light intensity does increase fairly linearly with arc length in GTAW at a fixed current and may be used as a feedback signal. It has been found, however [229], that at shorter arc lengths the intensity may also rise due to an increased concentration of metallic ions in the arc. Improved signals are obtained by selectively monitoring the arc emission spectrum and in particular the intense argon lines at wavelengths of 7504 Å, 7514 Å, 8006 Å and 8015 Å. This technique avoids problems with arc voltage fluctuations unrelated to arc length and may utilize a relatively simple sensor that is placed at a safe distance from the arc.

(c) *Penetration control*

Penetration control techniques have been developed mainly for situations where a full-penetration butt weld is required and welding is to be carried out from one side of the joint. These techniques may also be applied to critical root beads, for example in circumferential welding of pipe.

Backface sensing. It has been shown that the energy of certain wavelengths of radiation emitted from the back of the weld pool is a function of the weld bead area or width [230]. The radiation from the weld pool may be distinguished from that emitted by the parent metal by restricting the bandwidth of the sensor system to 450–550 nm, i.e. in the green part of the visible spectrum. In practice the radiation sensor may be a photodiode or phototransistor which is sensitive to visible radiation; this is mounted beneath the joint with a suitable filter and protective shield (figure 10.31), and the signal from the sensor is passed to the control unit where it is compared with a preset reference value. The error signal may be used to control a range of process parameters, the most common of which is pulse duration in pulsed GTAW. In this case the pulse is maintained at the high level until penetration is achieved, and the feedback signal reaches the reference level. When this occurs the current is reduced to its background level for a fixed time and the torch is indexed along the seam. The pulse duration varies to suit

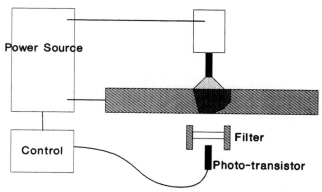

Figure 10.31 Principle of backface penetration control system.

the joint conditions, decreasing if there is a tendency for heat build-up and increasing if the thickness of the plate increases.

For industrial applications the radiation may be conducted to the optoelectronic sensor via a fibre optic cable as shown in figure 10.32. The technique is simple and reliable; it requires access to the backface of the weld but this is often easily achieved if optical fibres are used. Applications have included the butt welding of small-diameter tube to make transducers as well as the cir-cumferential joining of stainless steel tanker shells as shown in figure 10.32.

Figure 10.32 Backface penetration control sensor—fibre optic delivery of signal from backing gas supply cup to opto-electronic sensor.

The technique has also been investigated for GMAW using pulsed metal transfer and is capable of maintaining control of bead width with variations of root gap up to 2.5 mm and progressive variations in speed and thickness for butt welds in steel up to 3 mm thick [231,232]. It is possible to replace the optoelectronic sensor with a CCD camera and successful control systems based on this approach have been developed [233]. In plasma keyhole welding the process may also be controlled by monitoring the efflux plasma emerging from the rear of the keyhole with an optoelectronic sensor or video camera.

Front face light sensing. Penetration monitoring from the top face of the weld pool has recently become more practicable [234] with the development of a system which detects weld pool oscillations in the GTAW process. The technique involves the application of a short-duration high-current pulse (e.g. 100 A for 1−2 ms) which excites the pool and creates an oscillation. The oscillation frequency is observed using an optical sensor which picks up the reflected light fluctuations from the weld pool surface. It is found that an unpenetrated weld produces a higher frequency of oscillation than a penetrated weld and the difference is significant. By

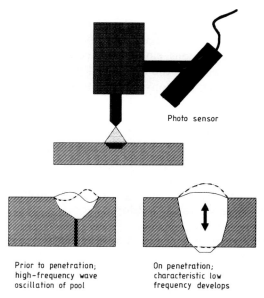

Photo sensor

Prior to penetration;
high−frequency wave
oscillation of pool

On penetration;
characteristic low
frequency develops

Figure 10.33 Principle of front face penetration control system.

analysing the frequency the reduction due to penetration may be detected and fed back to the controller. The arrangement and the proposed mechanism are shown in figure 10.33.

Voltage oscillation. The detection of forced weld pool oscillations by measurement of arc voltage has been successfully demonstrated [235] but this technique is complicated by the electrical noise and its application may be restricted.

Ultrasonic penetration control. Ultrasonic techniques for penetration control are less well developed than the corresponding seam tracking systems but control systems based on similar principles have been demonstrated. The combination of seam tracking and penetration control using a single sensor is attractive but the cost and complexity of the systems is likely to be high.

Pool depression. When the fully penetrated bead develops in GTAW the weld bead tends to drop and a depression is formed in the surface. If the torch height is fixed the corresponding increase in voltage may be taken as an indication of penetration [236]. Alternatively if an AVC system is being used the displacement of the torch may be used as a feedback signal. The feasibility of the technique as a continuous closed-loop control system has been demonstrated for welds in the downhand position but its application to other welding positions is likely to be restricted.

Front face vision. Direct observation of the weld pool shape may be used in conjunction with suitable models of the relationship between pool profile and penetration to predict the onset of full penetration. Although the system appears to be feasible complex image analysis facilities are required and mathematical models capable of accurately predicting the pool profile/penetration relationships have yet to be developed.

Radiographic control. Radiographic monitoring has been used for analysis of process performance in the slag-shielded processes such as SAW and ESW. The availability of high-quality real-time radiography and image processing techniques has led to the development of an on-line penetration and quality-control system [237]. The x-ray source is placed above the plate being welded and the image intensifier and camera are positioned below the work-piece. A computer control system is used to analyse the image and this provides appropriate corrections to the control system to ensure consistent penetration is maintained.

This approach is obviously complex and costly and is only likely to be used in special applications where no alternative technique is available and the overall cost is justified by quality constraints.

(d) *Quality control*

The seam tracking, arc length and penetration control systems described above are intended to compensate for relatively small variations in joint preparation, fit-up and material properties. Their application restores the welding performance to a predetermined level and reduces the likelihood of poor joint quality. Taking a pessimistic view; even if the weld is in the right place it may still contain metallurgical defects such as porosity and cracking. Ideally the control system should monitor the quality of the joint in real time and take corrective measures to ensure that the required standard is achieved.

Some steps towards the automatic control of quality have been taken but this area is still the subject of considerable research.

Bead geometry prediction. The finished bead geometry may itself be a quality criterion or there may be a clear relationship between geometry and secondary quality considerations. (For example, convex weld bead reinforcement may lead to stress intensification and subsequent cracking or high depth-to-width ratios may be responsible for solidification cracking.) If a sufficiently accurate model for the relationship between the weld variables and the geometry is obtained the process parameters may be adjusted on-line to produce the required geometry. Progress in the development of suitable mathematical models has recently been made but their application in process control is still at the research stage.

Thermographic sensing. Remote thermographic imaging of the weld pool has been found to be a practical [238,239] method of assessing the temperature profile of the joint in real time and enables the control of penetration, seam tracking and metallurgical characteristics of the weld. Using an expert system the observed temperature measurements may be related to the likelihood of defects and the probable mechanical properties of the joint. Sensing may be performed by a fibre optic and remote thermal imaging system or a thermal line scanner of the type shown in figure 10.34. The system is no more expensive than a laser stripe sensor but offers the possibility of much more comprehensive control.

Figure 10.34 Thermal line scanner—produces temperature profile of weld bead during welding. (Courtesy HGH, Ingéniérie Systèmes Infrarouges, France; Southern Scientific, UK.)

Hybrid systems. It is often possible to combine the information from several simple sensors to obtain a better indication of process performance and ensure more effective control. A torch displacement sensor when combined with through-arc measurements of voltage and current may, for example, be used to distinguish between wire feed slip and torch height variation. Increasing use of these hybrid systems combined with computer control should improve the ability to achieve true on-line quality control.

10.5   Summary and Implications

The range of control options for welding vary from the traditional open-loop manual systems based on welding procedures to complex closed-loop automatic techniques. Improved monitoring techniques and a wide range of sensors make it possible to measure process performance in both manual and automated applications and should enable more consistent weld quality to be achieved.

The application of the appropriate control approach should result in improved productivity and lower total cost. The application of all monitoring techniques and in particular computer-based systems is increasingly justified by economic and quality requirements.

The sophisticated adaptive control systems should not, however, be used to compensate for avoidable deficiencies in joint preparation and component quality; this will often prove costly and ineffective.

Of the automatic closed-loop systems described some of the simplest are the most effective.

11 Welding Automation and Robotics

11.1 Introduction

Many traditional welding processes are labour intensive and an analysis of welding costs shows that some 70 to 80% of the total cost may be accounted for by the labour element as shown in figure 11.1. Welding automation is a means of reducing the overall cost of the welding operation by replacing some or all of the manual effort with a mechanized system. The introduction of automation may, however, have much more significance than its primary effect on labour costs; in particular its influence on the following factors must be considered: safety and health; product quality and supply flexibility.

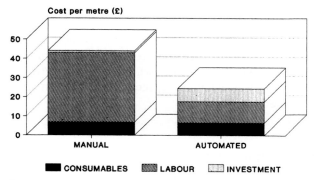

Figure 11.1 Comparative costs of butt weld in 20 mm thick steel (IG V butt, IG position, 60° V GMAW).

11.1.1   Safety and health

Most welding processes are potentially hazardous; they generate particulate fume, toxic gases, noise and a range of electromagnetic radiation which varies from ultraviolet radiation with arc processes to x-rays in electron beam welding. These hazards are well known and procedures for dealing with them have been established. The measures which must be taken to protect the welder and associated workers are, however, costly and may complicate the welding operation or involve the use of cumbersome protective clothing. There is also the risk of human error which may expose welders and those around them to unnecessary risk.

In addition to these process-related hazards there are risks associated with the application; such as welding in confined spaces, underwater or in radioactive environments. Automation offers a means of removing the operator from the process- and application-related hazards and in addition it offers the prospect of improving the control of the welding environment.

11.1.2   Product quality

Reproducible product quality may often be difficult to achieve with manual welding techniques, particularly when advanced materials and complex joint configurations are involved. Increasing the level of automation can significantly improve consistency, increase throughput and reduce the cost of inspection and rejection.

11.1.3   Supply flexibility

It is often easier to match output to demand with automatic systems than it is with labour-intensive operations. This is particularly true in welding situations where protracted training and qualification of welders may be required before an increase in output can be obtained.

11.2   Automation Options

Welding automation may vary from simple positioners to fully integrated systems. For clarity the various options will be discussed under the following headings:

(i) simple mechanization;
(ii) dedicated and special purpose automation;
(iii) robotic welding;
(iv) modular automation;
(v) programmable modular and CNC systems;
(vi) remote master–slave manipulators.

11.3   Simple Mechanization

The most common simple mechanization systems may be grouped
under the following headings:

(i) tractor systems;
(ii) positioners and manipulators.

11.3.1   Tractor systems

These are based on a simple electrically propelled tractor which
may be driven along the plate surface or may be mounted on a
track and driven by a rack and pinion. The welding head is
mounted on the tractor usually in some form of adjustable clamp.
Direct-mounted friction drive systems are usually considered
satisfactory for submerged arc welding but for GMAW and flux-
cored wire welding track-mounted gear-driven systems are preferred since they are less prone to slip. A typical system is shown in
figure 11.2.

The track is normally supplied in straight lengths for linear
seams but it is possible to obtain circular track rings for pipe
welding and integral annular tracks for circle cutting and welding.

The user may also adapt these devices to suit a particular
application using a wide range of standard accessories; these
include torch oscillation devices to allow positional GMAW welds
to be performed, trailers to carry ancillary equipment (e.g. wire
feed units), motorized cross slides and tactile seam-following
devices.

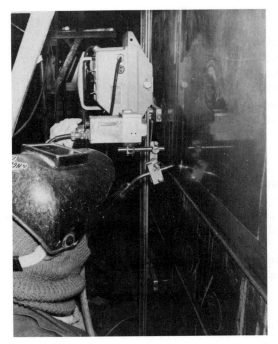

Figure 11.2 Vertical GMAW welding of stainless steel with a tractor system and oscillator. ('KAT' tractor, courtesy Maidel, UK.)

11.3.2 Applications of tractor-driven devices

The portability of the equipment makes this type of unit particularly suitable for welding applications on large fabrications such as marine structures, buildings, and storage tanks. A good example of this type of application is the use of a tractor and oscillator, of the type shown in figure 11.2, for the completion of some 5000 m of vertical and horizontal butt welds in the fabrication of austenitic stainless steel cell liners at the British Nuclear Fuels Ltd (BNFL) reprocessing plant at Sellafield [240]. The joint configuration was a square butt weld in 304L stainless steel, welded onto a plain carbon steel or stainless steel angle which was precast into the concrete of the cell walls (see figure 11.3). In this application pulsed transfer GMAW was used with a solid filler wire and electronic power sources equipped with synergic control. In this case the fabricator was able to produce consistent high-quality joints in a timescale which would have been impossible to meet using manual welding. A similar system was used [241] in the fabrication of 96 m of butt welds in a 24.4 m diameter crane tub in 35 mm thick BS 4360 50D material. It was estimated in this application that the cost saving was some £30 000 compared with vertical welding using the manual metal arc process.

Weld

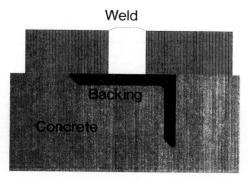

Figure 11.3 Weld joint configuration used in stainless steel cladding operation. (After Waering *et al* [240].)

These tractor-based systems are very adaptable and with a little ingenuity may be configured to suit a wide range of applications. An example of the use of a standard tractor system for a novel application is shown in figures 11.4. and 11.5. The application required the stainless steel nut insert to be welded at intervals into large carbon steel plates. A leak-proof fillet weld was required and skilled GTAW welders were not available. GMAW welding with a small-diameter stainless steel filler-wire and a helium/argon/CO_2 shielding gas was found to give acceptable bead profile but access and the high welding speed made it difficult to achieve consistent quality using manual techniques. The tractor and track system shown in figure 11.5 were therefore adapted as shown; a GMAW torch was substituted for the normal oxy-fuel cutting system, and an insulated peg was used to locate the assembly in the insert (and protect the thread from spatter damage). The total cost of the

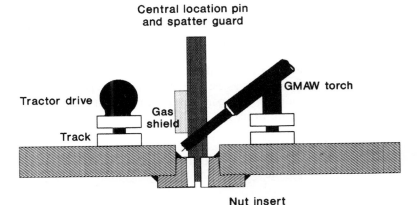

Figure 11.4   Fender nut welding system.

Figure 11.5   The use of a 'Hobo' circle-cutting tractor for welding.

mechanization system was under £1000 and a large number of high-quality joints were produced with an average welding time of 20 s. This example illustrates that although these systems are usually most suitable for long weld seams with simple geometry they may also be applied to much smaller joints. They are also more easily applied with the consumable electrode processes

(GMAW, FCAW, SAW) but special systems for GTAW and plasma welding are also available. This type of automation still requires constant supervision by a welder but the welder is removed from the immediate vicinity of the heat source and the exposure to fumes is reduced; the fatigue factor is also reduced.

11.3.3   Fixed welding stations

Simple rotary positioners (figure 11.6) and welding lathes may be used to move relatively small components under a fixed welding head or even a manually held torch. Using simple jigging this type of automation may easily be justified for relatively small batch sizes. It is particularly suitable for circular weld paths but fixed linear slides are also available for straight seams. Even the low-cost

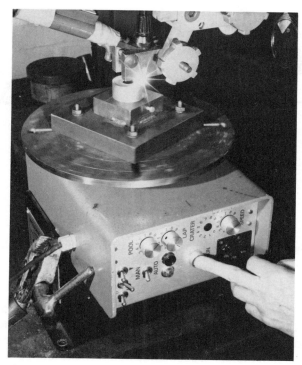

Figure 11.6   GTAW edge weld using a simple turntable.

units can include facilities for synchronization of power source switching and weld crater fill facilities. For larger workpieces column and boom positioners, motorized beams, roller beds and turntables are available [242]. Like the tractor systems these units are most suitable for simple geometric shapes and the consumable electrode processes, and they are adaptable to a range of applications limited only by the size and weight of the components being joined. The major advantage of this type of system is the ability to carry out welding in the downhand position which enables higher deposition rates and higher quality to be achieved. These advantages are most significant on heavier sections and can result in large cost savings in spite of the high initial cost of the positioning equipment. Typical configurations of these systems are illustrated in figure 11.7.

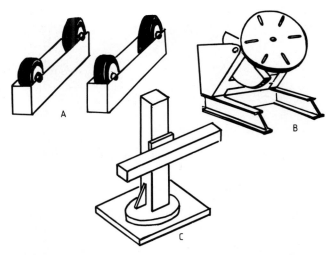

Figure 11.7 Heavy positioning equipment (not to scale): A, roller bed; B, tilting turntable; C, column and boom.

The most common application area for column and booms, roller beds and heavy turntables is for longitudinal and circumferential seams in the fabrication of pressure vessels and power generation plant. Very large units have also been used for making circumferential joints in submarine hull sections and power generation system drums as shown in figure 11.8.

Figure 11.8 Roller bed and column and boom being used to fabricate 40 tonne steam header for PWR power station. (Courtesy Aiton and Co. Ltd.)

11.4 Dedicated and Special-purpose Automation

11.4.1 Dedicated automation

Dedicated automation involves the design of a special welding system for a particular application and the resultant equipment may not be adaptable to changes in joint or component design. This type of automation is usually only justified for large production volumes of components with an extended design life.

Dedicated automation has traditionally been used for automotive components such as road wheels and exhaust systems with a wide range of welding processes including resistance spot, GTAW and GMAW.

The welding head is often only a single station in a multistation automation system which prepares the component for welding and also carries out finishing operations; in such cases a 'carousel'

design with a single load–unload station is often used. An example
of a system of this type, which has been specifically designed for
preparing automotive fuel injection components is shown in figure
11.9. Dedicated welding systems have also been employed where
lower production volumes and shorter product life cycles are envisaged but the welding environment is particularly hostile or the
quality of the end product is of primary importance. Examples of
this type of application are to be found in the nuclear industry,
both in processing of radioactive materials and the construction of
critical fabrications. An example of the type of equipment used for
this later application [243] is shown in figure 11.10; this was used
for GTA welding of advanced gas-cooled reactor (AGR) standpipe
joints on the Heysham II and Torness power station projects. The
use of costly dedicated automation with a sophisticated power

Figure 11.9 Dedicated welding system for welding fuel lines. (Courtesy
JPR Ltd, UK.)

Figure 11.10   Automatic welding system for AGR standpipe welds.

source† and control system was justified on the grounds of the unacceptability of defects and the repeatability of performance.

The need to purpose design the dedicated systems around a specific component usually makes the cost of such equipment high and many dedicated automation applications are now being tackled using the modular or the robotic approach.

† The power source used was a transistor series regulator with facilities for pulsed GTAW, programmed touch starting and arc length control. The welding head was also equipped with a vision system for remote monitoring of the weld area.

11.4.2   Special-purpose automation systems

Special-purpose automation has been developed for particular applications where similar joints are to be made on a range of component sizes. Some examples are simple seam welders, orbital welding systems and the GMAW stitch welder.

(a) Seam welders

The seam welding of sheet metal by the GTAW, plasma and GMAW processes to form simple cylinders or continuous strip has been a commonly recurring requirement. In order to cater for this type of application standard automated equipment which clamps the adjoining edges and moves a welding head along the seam has been developed. The equipment is adaptable to a range of material thicknesses and workpiece dimensions but dedicated to longitudinal seam welding.

(b) Orbital welding systems

The need to perform circumferential welds in pipe and tube fabrication applications is met by a range of orbital welding systems which include tube to tube heads, tube to tube plate and internal bore welders. These are usually portable systems which locate on or in the tube to be joined and rotate a GTAW head around the joint. Larger devices may be tractor-mounted on a circumferential track similar to the simple tractor systems described above, whilst the smaller systems utilize a *horseshoe* clamp arrangement (figure 11.11). Wire feeding and arc length control may be incorporated in the welding head and more sophisticated systems may allow the welding parameters to be changed progressively as the torch moves around the seam.

These systems are commonly used in power station construction for boiler tube joints and tube to tube plate welds. A good example of the productivity savings which can be achieved with these techniques when compared with manual welding is, however, the application of orbital welding techniques to the fabrication of more than 60 000 butt welds in stainless steel pipework at the BNFL reprocessing plant [244,245]. The application of orbital welding systems, together with improved pipe preparation and purging techniques, gave an improved first-time pass rate (from 50–60% to 87–90%) for each weld and more than halved the

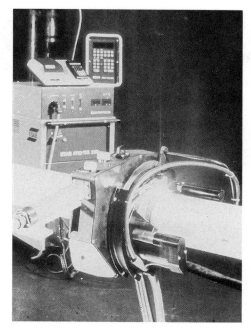

Figure 11.11 Orbital GTAW system with programmable power source and horseshoe-type head. (Courtesy ESAB.)

person hours per weld. The use of a pre-placed consumable socket [246] enabled simple square-edge pipe preparations to be used, provided joint alignment, avoided the use of a wire feed system, and allowed a single-pass welding procedure to be adopted. As in many applications of this type additional benefits were obtained by adapting the automation technique to suit the application.

(c) *GMAW stitch welder*

The GMAW stitch welder is a novel development in which a GMAW welding torch is mounted on a small motor-driven slide. The assembly is mounted in a head which automatically locates in the weld seam and mechanically fixes the torch height and angle (figure 11.12). The weld length and welding speed are preset by the operator (up to a maximum of 150 mm with this unit) and the welding process is initiated by a simple press button. The unit is ideal for producing consistent-size fillet welds and has been designed

Figure 11.12   Stitch welding equipment for GMAW.

for ease of use with the operator holding the unit in a vertical or horizontal orientation.

11.5   Robotic Welding

Industrial robots are not humanoid welders† but are defined by the British robot association as:

> An industrial robot is a reprogrammable device designed to both manipulate and transport parts, tools or specialized manufacturing implements through variable programmed motions for the performance of specific manufacturing tasks.

† The word robot was first used by Karel Čapek in his play *Rossum's Universal Robots* which was first published in 1920. The image of computer-controlled humanoids has persisted in science fiction and if anything the robot has been humanized further first by Asimov and more recently by Adams [247], 'Marvin the Paranoid Android'. Although industrial robots may be attributed human characteristics, particularly when they fail to operate in the intended manner, they are in fact simply programmable, and usually, computer-controlled actuators. Unfortunately the fictional image has tended to colour our perception of industrial robots and may have raised our expectations or influenced our judgement concerning their application.

In the case of welding robots the 'tools or specialized manufacturing implements' consist of welding heads, wire feed systems and tracking devices. And the processes for which robot welding systems are now available include GMAW, FCAW, SAW, GTAW, plasma, resistance spot, laser and NVEB welding.

The robot welding system consists of:

(i) a mechanical arm or manipulation system;
(ii) a welding package;
(iii) a control system.

11.5.1 Mechanical manipulation systems

A range of common configurations of manipulating system has evolved and these are illustrated in figure 11.13. The most common

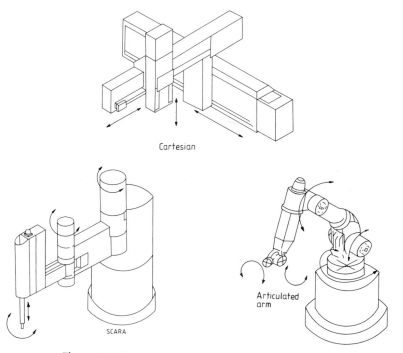

Cartesian

SCARA

Articulated arm

Figure 11.13 Typical welding robot configurations.

configuration for general-purpose welding robots is the articulated arm usually with six or more axes of movement, and a typical system is shown in figure 11.14. The advantage of the articulated arm is its flexibility and the ability to reach difficult access areas (it may be no coincidence that it has a similar configuration to the human arm).

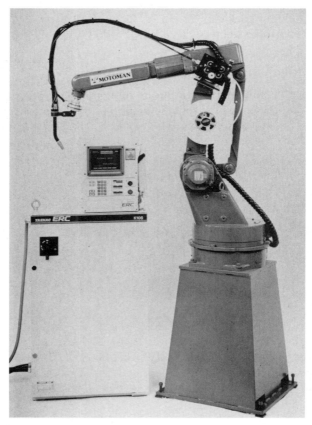

Figure 11.14 Normal articulated arm type welding robot. (Courtesy Torsteknik UK Ltd.)

The SCARA (Selective Compliance Assembly Robot Arm) con-figuration has traditionally been employed for assembly operations and has limited positional capabilities; it has, however, been used by some manufacturers as the basis of a simple, easily taught four to five axis machine for small batch production. A typical SCARA welding robot is shown in figure 11.15.

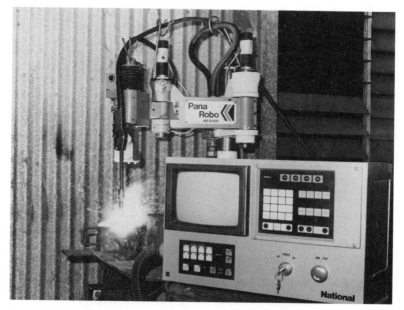

Figure 11.15 SCARA type robot being used for hardfacing application with FCAW.

Cartesian or gantry robots have been developed for very small high-precision applications and when very large operating envelopes are required.

These standard configurations may be adapted to particular applications and, for example, it is common to suspend articulated arm robots from overhead gantries for improved access, and it is possible to construct special robot configurations by rearrangement of the axes; whilst the final system may not appear to be related to the conventional welding robot it still has the essential facility for reprogrammability to suit different applications. A special-purpose robot of this type is shown in figure 11.16. In this case the seven-axis device is intended for making a range of saddle joints on pipe interconnections using the submerged arc process.

Another variation of the common configuration is the use of a linear slide to enable the whole robot to traverse the length of a component or a weld seam. Miniature portable and rail-mounted robots of this type have been developed for welding large structures.

Figure 11.16   Special-purpose programmable automation for pipe branch welds (SAW process).

(a) Drive systems

The arm may be driven by pneumatic, hydraulic or electrical actuators. Hydraulic power systems are suitable for applications requiring high load carrying capacity (above 35 kg [248]) and limited speed control and may be used for resistance spot welding but most fusion welding robots are now equipped with DC servo motor drives. Stepper motor drives have been used for small precision systems; these have the advantage of inherent feedback of output shaft position but suffer from lack of power.

11.5.2 The welding package

The welding package for robotic welding will obviously depend on the process being used but some important characteristics of these packages may be identified.

(a) *Welding packages for resistance welding*

For resistance welding the robot end effector needs to carry a portable resistance welding gun. It is important that the gun is robust to ensure repeatable operation but it must also be compact and manoeuvrable. Inevitably this leads to some compromise in design, and in order to carry the weight of the normal resistance welding head and the associated cables it is usually necessary to use a heavy-duty robot. The welding transformer may be separated from the welding head but this involves the use of heavy secondary cables and potential power losses. Some gun designs allow the electrode assembly to be changed during normal operation in order to access different parts of the fabrication.

Resistance welding is a *pick and place* type application: the robot places the welding head at the joint location; the electrodes close on the joint and the weld is made; the robot then moves the welding head to the next point and repeats the welding operation. The travel between points is carried at a high speed and neither the speed nor the absolute position in space need be accurately controlled during this motion. Since the resistance spot welding robot is not usually required to follow a seam it is not normally necessary to use joint location and following devices.

(b) *Welding packages for arc welding*

For arc welding applications a power source with facilities for remote control and output stabilization is required. The use of the electronic regulation systems and computer control discussed in Chapter 3 simplifies the control interface between the robot controller and the welding system as well as ensuring that repeatable performance can be achieved. In the GTAW system the robot simply needs to carry a fairly lightweight torch and cables whereas in GMAW and FCAW applications the filler wire must be fed to the welding head. Trouble-free wire feeding is essential to avoid system failure and it is desirable to mount the wire feeder at the rear of the robot arm with a fairly short conduit to conduct the wire from this feeder to the torch. Some systems even employ an auxiliary feeder immediately adjacent to the torch to ensure positive wire feeding and it is common to use large-capacity pay-off packs of low-curvature wire to improve feedability. Wire cutting and torch cleaning facilities must also be provided and in some cases interchangeable torch heads, are also used; these are stored in a carousel and may be automatically replaced during the robot cycle as shown in figure 11.17.

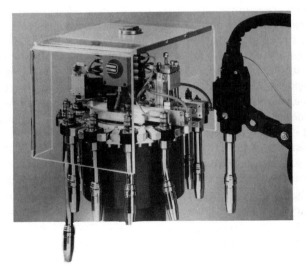

Figure 11.17   Torch head change system for robot. (BWS, courtesy Alexander Binzel, UK.)

(c)   *Welding packages for laser welding*
In laser welding applications the laser beam may be conducted along the robot arm to the workstation by a series of mirrors in the case of a CO_2 laser or by means of flexible fibre optic cables in the case of Nd:YAG lasers.

11.5.3 Robot control systems

The robot control system is required to:

   (i) control the position of the welding head;
   (ii) control the welding package;
   (iii) interface with auxiliary systems;
   (iv) interface with the operator;
   (v) provide program storage.

(a) *Control of position*

By driving three or more actuators simultaneously the end of the robot arm may be made to trace any path within its three-dimensional operating envelope. But to enable the position and velocity of the end effector to be controlled information concerning the position and rate of change of position of each actuator or axis must be obtained and processed. The position of the individual actuators is usually obtained from a *shaft encoder* which is attached to the output shaft of the drive motor. The information from these encoders is fed into the control system where it is recorded to enable the position of the arm to be duplicated. Very early systems recorded the encoder positions directly on magnetic tape and the information was *played back* through the servo motor controllers to duplicate the prerecorded path. The present generation of robots use microprocessor control systems and the positional information is usually compressed and stored in some form of non-volatile memory.

(b) *Control of welding functions*

The control system must coordinate the motion of the arm with the required welding functions. It must be able to initiate and terminate the welding operation in a controlled manner and should have facilities for setting the operating parameters for the process.

(c) *Interface with auxiliary functions*

The control system must be capable of receiving information from a number of auxiliary systems; for example, it must be able to respond to an instruction to start the welding operation and should be capable of checking various conditions, such as the presence of the workpiece in the welding jig and the closure of safety doors. It should also be capable of sending output signals to auxiliary

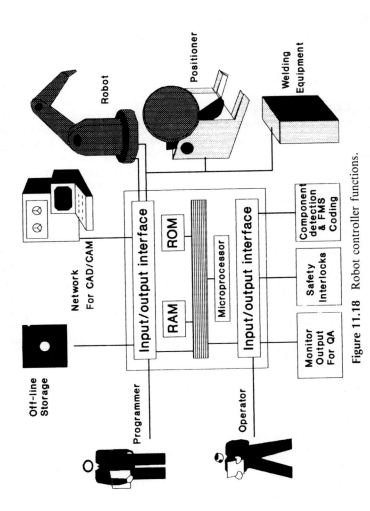

Figure 11.18 Robot controller functions.

systems; to initiate the motion of a work handling fixture for example. Most robot controllers are equipped with a large number of programmable input/output facilities of this type.

(d) Interface with the operator

There are several levels of operator interface with most robot systems. The simplest of these are the teaching/programming interface and the production/operator interface.

The programming interface allows the welding operation to be taught and checked whilst the production/operator interface may allow the selection of a particular preprogrammed *job* but often only allows the welding cycle to be initiated or terminated. The general structure of the robot controller and its interfaces is illustrated in figure 11.18.

11.5.4 Programming/teaching

The flexibility of robot systems relies on the ability to *teach* the system to perform a set of welding operations. The action of teaching or programming the robot is often discounted as being a relatively straightforward and rapid operation but in practice the creation of a satisfactory welding program may take a considerable time. The details of the teaching procedure depend on the robot but usually involve teaching the robot motion and operating instructions followed by editing the program and adding welding parameters. Two techniques are normally employed:

    (i) teach-by-doing, or playback, mode;
    (ii) point-to-point with interpolation.

(a) Teach-by-doing

The teach-by-doing, or playback, mode of programming was one of the earliest techniques to be used and involves the operator leading the robot through the weld path and continuously or periodically recording the position of the welding head. The SCARA system shown in figure 11.15 utilizes this approach; the robot is set in the 'TEACH' mode, the program is allocated a number by means of keys on the controller and the operator then moves the head manually to the first point to be taught, pressing a button at the rear of the arm to record the point. Successive points on the weld path are taught in a similar manner and weld start positions are identified by means of an alternative button. The program is terminated and stored in memory by means of another key on the controller. Welding parameters and input/output (I/O) sequences may then be edited into the program using the key pad and VDU screen on the controller. The whole procedure may then be run under computer control in the 'TRACE' mode (without welding) for final checking, and the 'PLAY' mode which performs the whole sequence in production may be initiated by a simple switch or push button connected to one of the I/O ports.

The procedure is very easy to learn and the programming operation is extremely rapid but for curved or complex shapes a large number of points need to be recorded. An alternative system uses a continuous path recording technique and a device mounted on the end of the torch to maintain the correct torch to workpiece distance.

One problem which is inherent in these systems is mechanical backlash. The encoders which are recording the position of the axes in the 'TEACH' mode do not take into account any distortion in the arm caused by the method of leading the torch to the work. This may lead to some inaccuracy in playback, although the rigidity of the small SCARA systems tends to minimize this problem.

(b)   *Point-to-point with interpolation*

This system is the approach most commonly used on fusion welding robots. A program identification number and the 'TEACH' mode are selected at the controller. The robot is then driven through a path in space using the normal actuators which are controlled from a manually operated pendant equipped with push buttons or a joystick (figure 11.19). At selected points the position is recorded, by pressing a key on the pendant. The mode of travel between points, the velocity and the choice of welding or non-welding operation are also possible using appropriate keys on the pendant. The travel mode choice is usually 'LINEAR, CIRCULAR or WEAVE' and the computer will interpolate an appropriate path based on the points which have been programmed. The taught program is stored at the end of the sequence and again it is possible to edit in further instructions, welding parameters or control sequences. This approach gives improved accuracy and facilities such as software-generated weave patterns but the programming process takes significantly longer than the

teach-by-doing method and it requires more care to avoid accidental collisions.

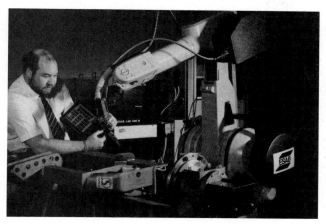

Figure 11.19   Hand-held control pendant for robot programming. (Courtesy ESAB.)

11.5.5   Program storage

The taught programs are stored in non-volatile memory (e.g.
battery-backed RAM) in the controller. It is also possible to
assemble a sequence of programs together into a batch file to
perform a particular job and to store regularly used sets of welding
parameters in a *library* file which may be called up during the main
welding program. †

For additional security, or to release storage space in the control
memory, the program, job, and library files may be transferred to
disc or magnetic tape.

11.5.6   Practical considerations

For successful implementation of robotic welding certain practical
considerations must be taken into account; these are safety, torch

---

† A *library* file may, for example, contain instructions to start welding, set
the current, set the voltage and the travel speed or to decrease current,
decrease voltage and turn off the welding system.

cleaning, jigging and positioning, work flow, component tolerances and joint design.

(a) *Safety*

Although robotic systems remove the operator from the immediate
vicinity of the welding process and reduce the risk from fume, arc
glare, noise and radiation they are in themselves a safety hazard.
The robot arm can travel at high velocity with considerable force
and the operator and associated staff must be prevented from
entering the operating envelope at any time when the unit is active.
This necessitates the use of mechanical guards and safety inter-
locks. In addition any automated system can operate at much
higher duty cycles than a manual operator and the total fume
generated during a shift will probably be much higher than that
produced by non-automated installations. It is therefore necessary
to provide adequate fume extraction.

(b) *Torch cleaning*

In automated GMAW operations the gas shield may become
ineffective if high levels of spatter accumulate on the end of the
nozzle. This problem may be minimized by the correct choice
of welding equipment and consumables but it will usually be
necessary to include a periodic torch-cleaning operation in the
robot program. Torch-cleaning stations are available for this
purpose.

(c) *Jigging and positioning*

The jigging system must locate the parts to be joined accurately,
be simple to load and unload and not present any undue obstruc-
tions to the movement of the robot arm. The capabilities of the
robot will often be extended by mounting the workpiece on a
programmable positioner which may be synchronized with the
movement of the robot arm to perform complex joint profiles. It
is also advantageous in many cases to suspend the robot from an
overhead gantry (figure 11.20).

(d) *Work flow*

A smooth flow of components to and from the robot system is
essential if the full capabilities of the unit are to be realized; this

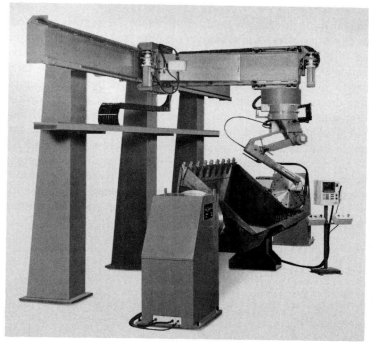

Figure 11.20 Robot system with programmable positioners and overhead suspension gantry. (Courtesy Torsteknik UK Ltd.)

will at the least mean careful scheduling of parts and may involve some investment in automatic delivery and discharge systems.

(e) Component tolerances

The positional repeatability of most fusion welding robots is of the order of 0.1 mm. The standard welding robot will *expect* the weld seam to be in exactly the same place and in the same condition with respect to gaps and misalignment as the original part which was used for programming. Deviation from these conditions may result in a defective weld and damage to the jigging. It is therefore important to establish the tolerance on fit-up and maintain the component dimensions within specified limits. In some cases this will involve improving control of component variability and may entail investment in pressing, cutting or machining equipment.

(f) *Joint design*

Careful joint design can be used to reduce the sensitivity to component tolerances, for example unsupported square butt joints and fillet welds in thin sheet may be replaced with lap joints to eliminate problems with gaps and give more tolerance to lateral movement. The fact that a robot can follow relatively complex three-dimensional paths should not be used as an incentive for using complex joint profiles; these not only make it more difficult to achieve the required quality, they often increase the cycle time. Simple joint profiles are usually the best. Accessability must also be considered when designing for robot welding; it may be impossible to weld a component completely without removing it from the jig and repositioning it. This can often be avoided by repositioning the welds to suit the robot. Some of these considerations are illustrated in figure 11.21.

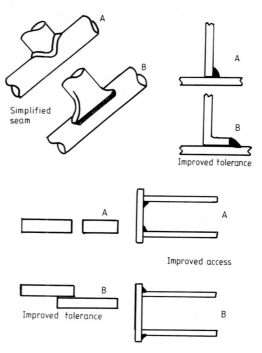

Figure 11.21 Modification of joint design to suit automation. In each case A is the original design and B the modified version.

11.5.7 Applications

(a) *Resistance spot welding*

Resistance spot welding robots form the largest single group of welding applications. Most of these are in the automotive industry where a group of robots can work simultaneously on a single body

Figure 11.22 Resistance spot welding on automotive production line. (Courtesy Fiat.)

shell as shown in figure 11.22. Articulated arm robots are most commonly used but there are examples of gantry systems being used for larger components, such as railway wagon side panels [249].

The robotic automation approach is appropriate in those applications where large production volumes are involved and model changes can be accommodated by reprogramming. The robot can also replace a physically difficult manual operation in an unpleasant environment. Although reasonable power and rigidity is required to carry the welding head and resist oscillation due to the rapid acceleration and shock loading, high levels of positional accuracy and precise control of linear velocity are not normally necessary.

Table 11.1   Robot applications

Applications	Material	Process	Robot	Reason	Reference
Ornamental gates	Hot-rolled mild steel	GMAW	One 5-axis artic. arm, one twin arm	Economic & increased output.	[1]
Hydraulic hoist parts	BS 4360 43A steel 10–12 mm.	GMAW	6-axis artic. arm	Reduced cost	[1]
Hospital beds	Square-section steel tube	GMAW	5-axis artic. arm	Reduced cost & improved quality	[1]
Dashboards	Aluminium	Spot	Modular cartesian	Increased output	[1]
Nuclear boiler panels	Steel	GMAW	Gantry-suspended artic. arms	Quality	[1]
Domestic boilers	Steel	GMAW	Twin arm	Skill shortage	[1]

continued

Table 11.1 (*Continued*)

Applications	Material	Process	Robot	Reason	Reference
Military storage system	3 mm thick Tenform XK-350	Spot	Cartesian	Economic	[1]
Vehicle chassis	Pressed steel	GMAW	16 artic. arm	Improved output & quality	[2]
Vehicle exhaust systems	Steel	GMAW	Artic. arm & manip. plus FMS	Increased output & improved quality	[3]
Turbine diaphragm	Stainless steel	Pulsed GTAW with filler	6-axis artic. arm	Quality & integration into CIM	[4]
Space shuttle engines	Nickel-based, cobalt-based, Ti alloy and stainless steel.	Pulsed GTAW	Various	Quality	[5]
Car bodies	Low-carbon steel	Spot	6-axis artic. arm	Output & quality	[6]
Engine support	Low-carbon steel	Laser	6-axis artic. arm	Output & quality	[6]

[1] Various authors *Metal Construction* **16**(4) April 1984

[2] Hanley TE Putting robotic welding into practice—the Land Rover chassis line *Metal Construction* **18**(4) April 1986

[3] Anon. Developments in arc welding automation *Metal Construction* **19**(9) September 1987

[4] Anon. Turbine manufacturer makes move to robotics *Welding J.* November 1986

[5] Flanigan L Factors influencing the design and selection of GTAW robotic welding machines for space shuttle main engines *Welding J.* November 1986

[6] Anon. Tipo—The product of the system *Publication* 10067-1/88 (Fiat Auto S.p.A. Turin)

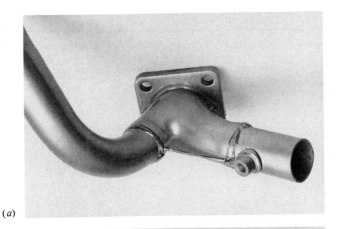

(*a*)

(*b*)

(*c*)

Figure 11.23 Typical robot-welded components: A and B, vehicle exhaust parts; C, bicycle frame. (Courtesy Torsteknik UK Ltd.)

(b) *Seam welding arc processes*

Continuous seam welding using GMAW, FCAW, GTAW and plasma processes has been applied in a wide range of applications from car exhausts to space shuttle components as shown in table 11.1. The predominant applications have been concerned with the fabrication of thin-sheet-steel pressings using the GMAW process and some typical examples are shown in figure 11.23.

Seam welding applications require accurate positioning of the welding torch and precise control of the travel speed during the welding operation. Programming time is usually significant and component tolerances and jigging are important. For these reasons reasonable batch sizes are expected to justify the investment in robotic automation.

Figure 11.24 Articulated arm robot with YAG laser and fibre optic beam delivery system. (Courtesy Kuka and Lumonics.)

(c) *Robotic power beam welding*

The use of robots for laser and electron beam welding applications is at a very early stage of development. Trials of robotic systems for out-of-vacuum NVEB are taking place whilst industrial applications of laser welding are in progress; for example in the welding of engine support frames for the Fiat Tipo [250].

Specially designed robots have been produced for laser welding with either mirror or fibre optic beam delivery systems incorporated in the arm, as shown in figure 11.24.

11.6 Modular Automation

As the cost and limitations of fully dedicated automation have become clearer attempts have been made to design lower-cost flexible systems which may be configured to suit a range of applications. This has led to the development of modular automation systems.

Modular systems use a kit of common mechanical components which can be assembled into any configuration to suit the application. The modules consist of support beams, slides, carriages, pivots, linear and rotational drives, torch holders and control systems as shown in figure 11.25. The advantages of this approach are rapid system design, simplicity and low cost.

Figure 11.25 Kit of parts for flexible modular automation system. (Courtest SAF.)

A welding automation system, such as the one shown in figure 11.26, may be developed very quickly using this approach and costly design studies are eliminated. The standard parts are easily interchangeable in the event of damage or breakdown and the total cost of the installation is kept to a minimum. The system may be used for most of the common arc welding processes and is particularly suitable for GMAW, GTAW and plasma welding.

Control of position and operating sequences and interface with the welding equipment is provided by a simple electronic logic system. The capabilities of the system may be extended by using a standard industrial programmable controller.

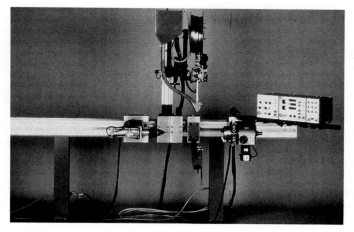

Figure 11.26 Typical welding system based on modular kit. (Courtesy SAF.)

11.7   Programmable Control

The control systems approach used for robotic welding may be applied to special-purpose, dedicated and modular automation systems to improve their flexibility.

This approach has been adopted, for example, on orbital GTAW equipment to allow complex welding procedures to be developed and stored for specific joints. In this case a computer control system is usually incorporated in the power source or as an interface between the welding power source and the welding head. A typical system is shown in figure 11.27.

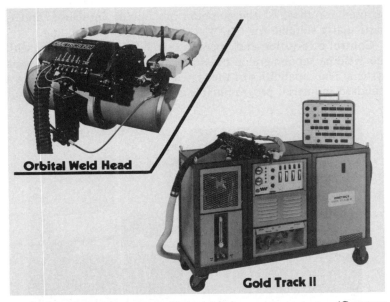

Figure 11.27 Computer-controlled orbital GTAW system. (Courtesy Dimetrics and Newtech UK.)

For more general applications with dedicated and modular automation, flexible computer control systems have been developed [251]. These systems have the following facilities:

(i) positional control;
(ii) weld process control;
(iii) cell management.

The positional control system controls the position and velocity of a number of axes, either rotational or linear, to enable three-dimensional trajectories to be followed. Both the welding head and component position may be controlled. The weld process control system may control the welding parameters directly or more commonly via an interface with an *intelligent* welding power source (see Chapter 3). Cell management activities concern the communications between the welding system and the external production environment. The control should be able to actuate component delivery and discharge systems, provide job status and quality information.

The combination of computer control and modular mechanical design offers an alternative to the normal robotic approach.

11.8 Remote-control Slave and Automatic Systems

Remote-control welding devices are used in particularly hazardous environments. They may take the form of a master–slave manipulator or a fully automatic system with remote monitoring.

11.8.1 Master–slave manipulators (MSM)

This type of system involves the use of a multi-axis positioner which is positioned and controlled by a remote manual operator.

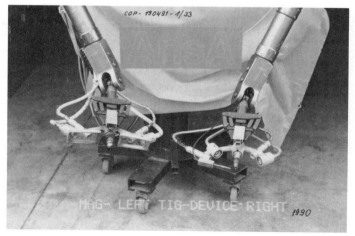

Figure 11.28 Remote-controlled master-slave manipulator (MSM) arm. (Courtesy of ANSA.)

These devices have been used in the nuclear industry for the manipulation of radioactive components and in hyperbaric applications for positioning components or welding inside a high-pressure chamber.

These systems are usually specially built to meet the specific application requirements; although some general-purpose arms are available† these are not normally designed for welding applications and the repeatability, positional accuracy and load carrying performance must be evaluated carefully. Figure 11.28 shows a typical MSM arm: welding systems of this type have been specially developed for nuclear applications.

11.8.2 Fully automatic remote welding systems

The use of fully automatic systems for remote welding applications reduces the possibility of manual error and should improve

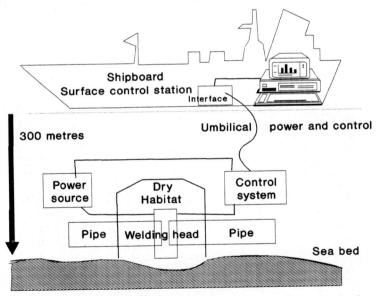

Figure 11.29 Automatic hyperbaric GTAW system for orbital pipe welding.

† Special-purpose remotely controlled arms are normally designed for underwater applications or bomb detection and disposal.

(a)

(b)

Figure 11.30 Hyperbaric pipe welding system. (a) Pipe welding head in foreground, habitat in background. (b) Shipboard control station. (Courtesy of Norsk Hydro, Norway.)

repeatability. Systems have been developed for deep-water hyperbaric applications [252,253] in which welding is carried out at depths of up to 360 m in a dry hyperbaric chamber filled with helium-rich gas. The welding head is an orbital GTAW system and an inverter-based electronic power source under computer control is situated within a service container adjacent to the welding enclosure (i.e. on the sea bed). The process is controlled from a shipboard computer interface which allows the welding parameters to be 'downloaded' to the welding system. The welding sequence is performed automatically with visual (CCTV) and process parameter monitoring being carried out by the remote operator on the surface. A diagrammatic overview of the system is shown in figure 11.29. and a photograph of the main hyperbaric components is shown in figure 11.30. This system has been proved technically and economically in production on more than 30 pipe welds in pipe diameters from 250 to 900 mm at depths of 110 to 220 m.

Dedicated automatic systems have also been developed for nuclear applications [254,255], for example to weld circumferential seams in beryllium containers; a computer-controlled system incorporating a transistor series regulator GMAW power source was used in this case.

11.9   Advances in Welding Automation

Advances have been made in the development of systems and applications for welding automation, in particular:

(i) adaptive control;
(ii) flexible manufacturing;
(iii) simulation and off-line programming;
(iv) integrated systems.

Adaptive† control techniques may be used to improve the tolerance of automation systems to normal variations in component dimensions, joint position and material characteristics. These techniques should not be regarded as the first or only solution to prob-

† See definitions of adaptive control in Chapter 10.

lems of this type; redesign of the joint or improved component preparation are often more cost effective. Major developments have, however, been made in adaptive control and a large number of alternative options are available, as already discussed in Chapter 10.

Flexible manufacturing systems (FMS) are used in many non-welding applications†; they enable reductions to be made in work in progress by automatically scheduling the production operations to suit component availability and demand. A simple but effective example of this system is the Autotech flexible welding system [256]. In this system a conveyor is used to transport standard 750 mm × 1200 mm pallets containing the components to be welded from the loading station to the robot. Each pallet has jigging suitable for a specific component mounted on it and is identified by a unique arrangement of five plugs on its side. When the pallet enters the robot station the position of the plugs is read by proximity switches and fed to the robot controller which selects the

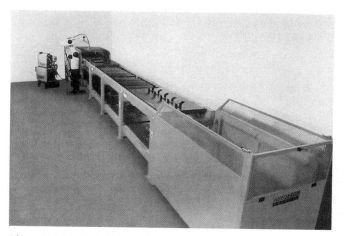

Figure 11.31 Flexible welding system. (Courtesy Autotech Robotics.)

† Flexible manufacturing systems are systems which are able to accept a variety of components or tasks, often in a random order, and are able to identify the component and automatically adapt the manufacturing operations to produce the required end result.

appropriate welding program. The pallets may be loaded onto the system in any sequence and the throughput may be varied to suit the availability of parts. A typical flexible welding system of this type is shown in figure 11.31.

This approach has been extended by the use of AGVs (automatic guided vehicles) to transport the workpiece and alternative methods of component identification [257]. The use of typical FMS systems in the fabrication of heavy construction equipment is reported to have resulted in cost savings of 28% [258]. The FMS approach is not restricted to robotic welding and may be applied to dedicated and modular systems which are equipped with suitable programmable controllers.

11.9.1   Simulation and off line programming

Programming of robotic systems may take a significant amount of time and this results in lost production. Off-line programming allows the robot program to be developed on a remote work-station and transferred to the robot controller almost instantaneously. Various systems are available for off-line programming (e.g. GRASP and PLACE, ROBCAD) but these usually share the following characteristics:

(i) A graphic and kinematic model of the robot is stored as a library file in a computer (either a mainframe or a personal computer depending on the system).

(ii) The workpiece and joint description may be loaded into the program and the welding operation may be simulated using the selected robot.

(iii) The completed simulation may be translated into the *real* robot's control language and transmitted to the system via a data link.

The BYG GRASP system is a typical simulation and off-line programming package; it offers fast and realistic solid model simulations of the application enabling the proposed installation to be fully evaluated and optimized without incurring production downtime (figure 11.32). CAD files of workpieces and fixtures may be read into the package which is equipped with a large library of robot definitions. The performance of the robot cell may be tested

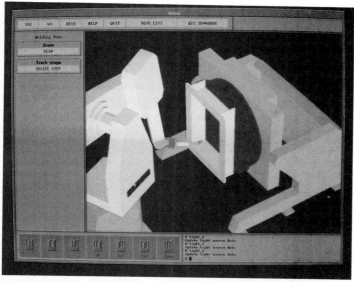

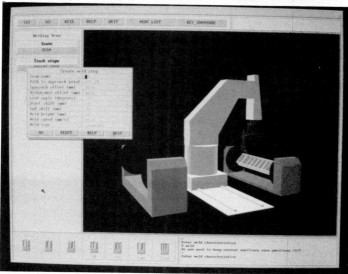

Figure 11.32 'GRASP' three-dimensional graphical simulation package with off-line programming capability. Screens show solid model simulation of typical welding applications. (Courtesy BYG Systems Ltd.)

and potential problems (e.g. collisions and access limitations) are readily identified and displayed on the monitor. The system also offers the facility to generate CAD drawings of the final work cell and transfer the simulated program to the control system using the native language of the appropriate robot.

An interesting example of the off-line programming approach is its application to shipbuilding. The system shown in figure 11.33 consists of a portal frame from which an articulated arm robot is suspended; the frame is lowered into the work area and an initial joint-locating program checks the orientation of the system and corrects the datum settings. Off-line programming is used to prepare the program and this is downloaded to the controller so that once the unit has established its exact position it may carry out the prescribed welds.

Figure 11.33 Robot system for shipbuilding (gantry-suspended articulated arm with feedback control and off-line programming facilities). (Courtesy Cloos.)

A similar system has been used in the Odense shipyard in Denmark with a Hirobo NC programmable robot. The NC programs created on a personal computer are in this case transferred to the robot controller via a plug in bubble memory and the robot is only out of production for 30 s [259]. The system has also been linked to a simulation package and the shipyard's CAD facility to enable more efficient off-line programming to be accomplished in the future.

11.9.2 Integrated automation systems

Many of the automatic welding systems of the type described above have been developed into integrated systems in which the welding cell is self-contained but linked to other manufacturing processes by a data communication network. Several manufacturers now offer robotic welding 'cells' configured for a particular range of applications and supplied complete with all the necessary services including safety screens and fume extraction. A typical example is shown in figure 11.34.

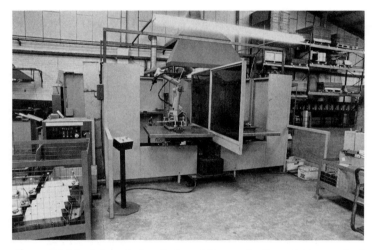

Figure 11.34 Integrated robot welding cell (two-station swivel table). (Courtesy G M Fanuc Robotics (UK) Ltd.)

Alternatively, dedicated or modular systems with computer control may be used to construct the basic cell (figure 11.35). The integrated welding system approach may also be used with computer-controlled modular automation or dedicated systems. Designs of this type have been evaluated for the production of high-accuracy GTAW welds on a range of components of similar design which span a range of sizes.

To accommodate the size range the component pallets are equipped with radio-frequency identification tags (Eureka tags) which communicate with the controller and reset the system for the component to be welded. To ensure accurate torch positioning in this type of installation the electrode position may be measured before each run by moving it between a pair of crossed light beams (this has been shown to be capable of ensuring positional repeatability to within ± 0.025 mm). The overall management of the system can be handled by a proprietary CNC controller which may also act as an interface for data communication to an associated cluster of preparation and finishing stations.

Systems of this type may be regarded as the equivalent of a machining centre and are easily integrated into a totally integrated production facility. It is clear that this approach can ensure greatly improved quality, high productivity and flexibility of supply.

Figure 11.35 Computer-controlled integrated welding system (three rectilinear axes plus two circular axes). (Acuweld 500, courtesy Huntingdon Fusion Techniques.)

11.10 Evaluation and Justification of Automated Welding

Systematic appraisal of the possibilities for automation has been studied by many authors [260–262] but the main steps involved are shown in figure 11.36 and may be listed as follows:

(i) define objectives;
(ii) assess the application;
(iii) examine the options;
(iv) evaluate alternatives;
(v) evaluate implications.

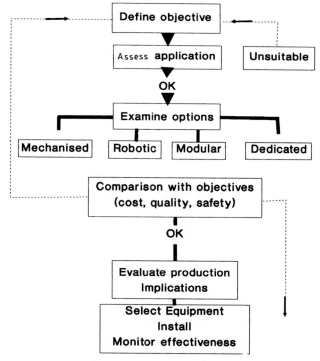

Figure 11.36 Automation decision network.

11.10.1 Objectives

The importance of clear objectives cannot be over-emphasized: failure to define the true reasons for seeking to use automation may result in considerable waste of effort. In general the objectives are likely to be based on the economic, quality, safety or supply flexibility requirements outlined in the introduction but in some cases the true objective may be more indirect; comments such as 'to demonstrate the company's involvement in high technology', or 'to worry the competition' or even 'because the managing director believes we need a robot' are not uncommon. These management aims are equally acceptable objectives but their implications must be considered when the success of the installation is being measured.

11.10.2 The application

The suitability of the application can be assessed in terms of the product life-cycle, the accuracy of joint preparation, the possibility of redesign [263] and the possibility of using a process which may be automated.

11.10.3 The options

The options available have been detailed above and in any application some may be readily eliminated as inappropriate at an early stage, but those with any possibility of success should be evaluated more carefully. This evaluation may involve process feasibility trials and examination of similar production systems but it will also involve a financial justification.

11.10.4 Financial justification

The type of economic analysis which is used will often be a matter of company policy and may involve an evaluation of return on investment or discounted cash flow, but in practice most assessments are based on the simple payback period†. In the case of structural steel fabrication, the labour cost often accounts for the major part of the total cost of welding. The primary cost savings

† In Japan a recent survey showed that 66.8% of companies used payback periods of three to five years as justification criteria whilst 10% used *net present value* techniques and only 6.5% used *internal rate of return* methods (Huang PY *et al* 1989 *Manuf. Rev.* **2** (3) Sept).

are therefore associated with improvements in the operating factor for the process (the ratio of effective to non-effective time) and the consequent reduction in labour cost.

For example, a manual GMAW operator may achieve an operating factor of 15–20% whereas with a tractor-mounted system an operating factor of 30–40% may be possible and fully automated systems are likely to achieve 80–90%. Secondary cost savings can also be expected from improved control of weld size; which in turn saves time, reduces consumable costs and improves control of operating technique; which produces more consistent quality, reduces defect levels and decreases repair costs.

A preliminary evaluation of the economic factors associated

Table 11.2   Example of costing spreadsheet output showing the influence of £20 000 investment in welding automation on the cost of making a butt weld in steel with GMAW, 1.0 mm diameter filler wire. The left-hand column shows the original cost for manual GMAW, and the right-hand column shows the estimated effect of automation on the capital cost, operating factor (arc on time) and weld quality (rejection rate). The totals show the reduced cost per weld and increased productivity to be expected.

Deposition rate	kg h^{-1}	3.55	3.55
Costs for flux or gas per m^3 or kg	£	3.10	3.10
Costs per 1000 electrodes or kg wire	£	0.77	0.77
Labour (without equipment) per hour	£	14.08	14.08
Number of hours per annum	h	1500.00	1500.00
Process arc-on time	%	28.00	50.00
Amount of work under survey	%	100.00	100.00
Rejection rate	%	5.00	1.00
Total investments	k£	2.535	22.535
Interest	%	8.00	8.00
Depreciation period	a	5.00	5.00
Total costs per annum	k£	28.788	36.500
Deposited weldmetal per annum	kg	1416.45	2635.88
Weight per metre	kg	0.33	0.33
Total costs per kg weldmetal	£	20.32	13.85
Cutting costs per metre	£	0.00	0.00
Costs for this process per metre	£	6.68	4.55
Cumulative costs for this weld	£	6.68	4.55

with the automation of a given application is straightforward, particularly if one of the many commercial weld-costing software packages† is used. Examples of simple cost comparisons made with the NIL COSTCOMP software are shown in table 11.2 and figure 11.37: a comparison of manual and mechanized welding approaches to one of the most common weld configurations is shown in table 11.2. The cost of the simple tractor involved would be recovered after only 315 m of weld had been performed. In the same way the high cost of a dedicated CNC unit may be offset by improvements in operating factor and improvements in quality (reduction in reject and repair rate).

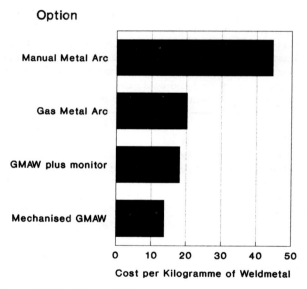

Figure 11.37   Cost analysis for 6 mm fillet weld in steel with various cost-reduction options (data from COSTCOMP weld-costing program).

---

† For example, the UK Welding Institute's 'WELDCOST' programme, The Netherlands Welding Institute's 'NIL COSTCOMP' or any suitably configured spreadsheet software.

11.10.5   The expectations

Assessments of the implications of introducing welding automation are often inaccurate, and surveys [264–267] carried out before and after the introduction of automation illustrate this point. It is worth considering some of the expectations and the resultant assessments prior to investment in automation.

(a)   *Social implications*

Social implications, and in particular shop-floor opposition, are often expected to be barriers to the introduction of automation; in fact whilst 31% of users in one survey believed this to be the case prior to the purchase of automated systems only 2% actually experienced any problems. In fact recent reports suggest that most workers prefer working with computer-controlled equipment and achieve additional status amongst their peers for doing so. It is also clear from these reports that insufficient attention is given to involving workers in the decision to automate and a review of successful applications shows that successful implementation of automation does depend on the commitment and ingenuity of the welding engineer and the operator.

(b)   *Production implications*

It is also found that most robot users underestimate the degree of after-sales support needed, the maintenance requirements and the development costs involved in bringing the application to full production. One of the most serious problems encountered is often the unsuitability of component tolerances and this may have major cost implications if preparation equipment needs to be replaced or operating procedures need to be changed.

11.11   Summary

The wide range of automation systems described in this chapter are capable of answering the need for:

(i) reduced production costs;
(ii) improved safety and health;

(iii) consistent product quality;
(iv) improved supply flexibility.

An evaluation of the potential benefits to be gained from welding automation is always worthwhile. In the unlikely event that none of the options listed above prove viable it is often possible to use the information gained in the analysis of automation requirements to improve the manual welding operation.

Appendix 1
Welding processes classification

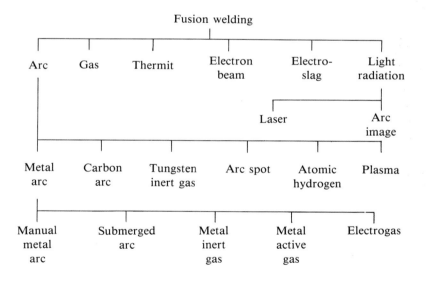

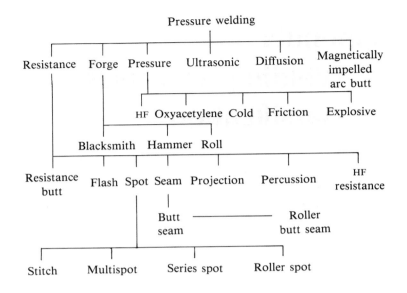

Appendix 2
SMAW electrode
classification

(reproduced by kind permission of *Welding and Metal Fabrication*).

BS 639

The standard identifies electrodes by a series of letters and digits that indicate the mechanical properties, formulation type and applications. The full electrode classification consists of a compulsory and optional parts as shown below.

Compulsory

E	Electrode for MMA
xx	Tensile strength range, minimum yield strength
x	First digit for elongation and impact strength
x	Second digit indicating temperature for minimum average impact strength of 47 J
NN	Letters indicating coating type

Optional

xxx	Three-digit indication of metal recovery
x	Welding position
x	Current and voltage requirements
H	Hydrogen-controlled electrode

This classification is quite complex and it is recommended that the standard should be consulted for full details. The following notes give an indication of the typical electrode specifications.

(BS 639)

Tensile Strength

Designation	Strength (N mm^{-2})	Minimum yield strength (N mm^{-2})
43	430–550	330
51	510–650	360

First digit: elongation and impact

Digit	Minimum elongation (%)		Impact test temperature ($^\circ$C) for impact value of 28 J
	E43	E51	
0	20	18	Not specified
1	20	18	+ 20
2	22	18	0
3	24	20	− 20
4	24	20	− 30
5	24	20	− 40

Second digit: toughness

Digit	Temperature ($^\circ$C)
0	Not specified
1	+ 20
2	0
3	− 20
4	− 30
5	− 40
6	− 50
7	− 60
8	− 70

Coating type

B	Basic	$CaCO_3$, CaF_2
BB	Basic iron powder	$CaCO_3$, CaF_2 + iron powder
C	Cellulosic	Organic
O	Oxidizing	FeO
R	Medium-coated rutile	TiO_2
RR	Heavy-coated rutile	TiO_2, $CaCO_3$
S	Others	

(BS 639)

Welding position
1   All positions
2   All except vertically down
3   Flat and horizontal
4   Flat
5   Flat, vertically down, HV fillets
9   Any combination not classified in 1–5

Current and polarity

Digit	DC	AC
0	As recommended by manufacturer	Not suitable
1	+ or −	50 minimum OCV
2	−	50 minimum OCV
3	+	50 minimum OCV
4	+ or −	70 minimum OCV
5	−	70 minimum OCV
6	+	70 minimum OCV
7	+ or −	80 minimum OCV
8	−	80 minimum OCV
9	+	80 minimum OCV

Typical example
The classification for an all positional basic (with iron powder) hydrogen-controlled electrode with a minimum UTS of $510\,N\,mm^{-2}$ and yield strength over $450\,N\,mm^{-2}$ and impact properties better than 28 J at $-40\,^{\circ}C$ and 47 J between -40 and $-50\,^{\circ}C$ would be:

E   51   5   5   BB   160   2   8   H

AWS Standard A5.1
The AWS Standard is less complicated, consisting of a prefix
E followed by only two sets of digits

E	Electrode
xxx	Two or three digits indicating the tensile strength in 10 000 psi units.
xx	Two digits indicating the coating type and application as follows:
Exx10	Cellulosic, DC + ve. Deep penetration, all position
Exx11	As Exx 10 but usable on AC
Exx12	Rutile, AC/DC, flat and HV positions
Exx13	Rutile, AC/DC, all positions
Exx14	Iron powder rutile, high speed
Exx15	Basic low-hydrogen, DC + ve, all positions
Exx16	Basic low-hydrogen, AC/DC, all positions
Exx18	As Exx16, with iron powder for improved recovery
Exx20	Mineral oxide/silicate, for flat and HV positions
Exx24	Similar to Exx12 but with iron powder, flat and HV
Exx27	As Exx20 but with iron powder
Exx28	Low-hydrogen basic with 50% iron powder, flat and HV

A suffix (e.g. A1, B2) may be added to indicate the chemical composition
of low-alloy-steel weld metal.

The electrode classified above as E 51 5 5 BB 160 2 8 H
would be classified as E7018 according to AWS A5.1.

Appendix 3

Appendix 3a
Burn of characteristics
GMAW Plain carbon steel

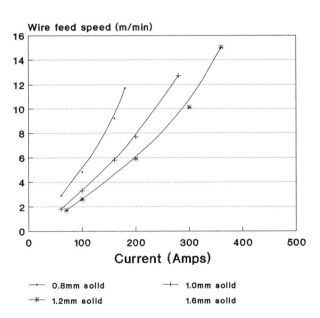

Wire feed speed (m/min)

Current (Amps)

- 0.8mm solid
- 1.2mm solid
- 1.0mm solid
- 1.6mm solid

Appendix 3b
Burn of characteristics
GMAW Stainless steel

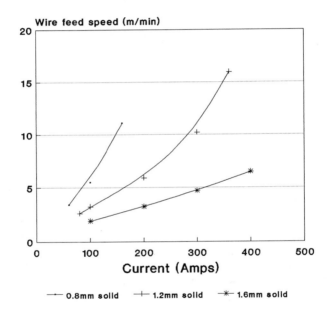

Appendix 3c
Burn of characteristics
FCAW Plain carbon steel

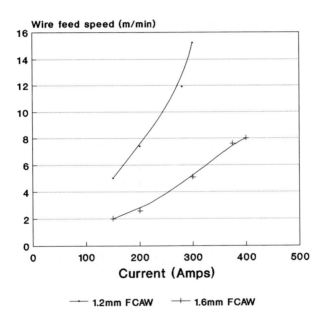

Wire feed speed (m/min)

Current (Amps)

── 1.2mm FCAW   ─+─ 1.6mm FCAW

Appendix 3d
Influence of electrode extension
FCAW Plain carbon steel

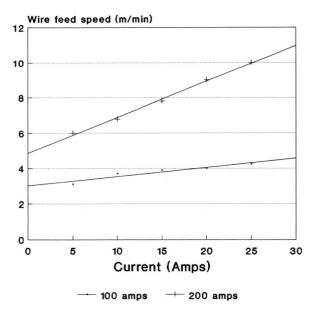

1.2mm Flux cored wire

Appendix 4

(reproduced by kind permission of *Welding & Metal Fabrication*).

The AWS classification of flux and cored wires for carbon manganese steels (A5.20) and low-alloy steel electrodes (A5.29).

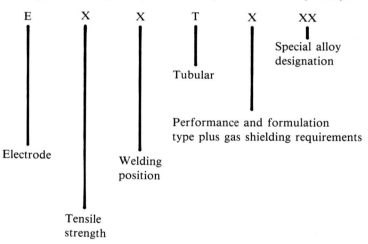

Minimum tensile strength in 10 000 psi units, normally:
6—60 000 psi
7—70 000 psi
8—80 000 psi, but alloy grades
11—110 000 psi, up to 12—120 000 psi

Welding position:
0 = flat and HV
1 = all positions

Performance and formulation details:

Digit	Gas	Details
EXXT-1	CO_2	Rutile-type, smooth running, general-purpose electrode
EXXT-2	CO_2	As T-1 but high Mn/Si ratio for single-pass HV fillets
EXXT-3	None	Self-shielded, thin-material, spray transfer
EXXT-4	None	Self-shielded, low-penetration, general-purpose, HV or 1G, globular
EXXT-5	CO_2	Basic type, high-toughness, hydrogen-controlled, globular transfer
EXXT-6	None	Self-shielded, deep-penetration, good mechanical properties
EXXT-7	None	Self-shielded, high-deposition-rate
EXXT-8	None	Self-shielded, positional and improved toughness
EXXT-10	None	Self-shielded high-speed, single-pass
EXXT-11	None	Self-shielded, general-purpose
EXXT-G & T-GS	As required	Special grades, e.g. metal-cored

The British Standard classification for carbon–manganese steel tubular cored welding electrodes (BS 7084.1989).

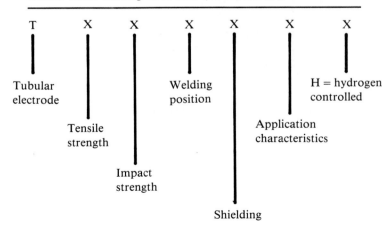

(BS 7084.1989)

Tensile properties:

Number	Tensile strength (N mm^{-2})	Yield stress (N mm^{-2})	Elongation, minimum (%)
4	430–550	330	20
5	510–650	360	18

Impact properties (temperature for minimum of 47 J):

Number	Temperature (°C)
—	None
0	0
2	− 20
3	− 30
4	− 40
5	− 50
6	− 60
7	− 70

Shielding: G = Gas; N = No Gas

Welding position:
  0   Flat and horizontal/vertical
  1   All positional
  2   All positions except vertically down
  9   Any position or combination of positions not classified in 0–2

(BS 7084.1989)

Applications and characteristics:

Letter	Gas	Details
R	CO_2	Spray transfer flat and HV position, usually rutile-type
P	Argon-based or CO_2	Spray transfer all positions, usually rutile formulation
B	CO_2 or argon-based	Basic formulation for good mechanical properties
M	Argon-based or CO_2	Metal cored—total non-metallic materials less than 1% of electrode mass
U	None	High deposition rate in flat and HV positions
V	None	Single-run high travel speeds in flat and HV positions
W	None	Multirun flat and HV positions
X	None	High metal powder, low-flux, all positional
Y	None	High-flux low-metal powder, all positions
S	—	Other types not specified above

Hydrogen controlled: the H suffix is used when diffusible hydrogen levels of less than 15 ml/100 g are obtained (when tested in accordance with BS 6693 Pt 5). The manufacturer must specify the maximum diffusible hydrogen expected (i.e. 5 ml/100 g, 10 ml/100 g, 15 ml/100 g), and the operating parameters at which these values are achieved.

Example: a CO_2-shielded, basic, controlled hydrogen electrode with all positional capabilities may be specified as:

$$T \quad 5 \quad 5 \quad 1 \quad G \quad B \quad H$$

NB: this summary is reproduced by permission of BSI.
For full details, including a full description of the consumables, see British Standard BS 7084.1989.

Appendix 5
Flux-cored wire for surfacing and wear resistance

Consumable type	Features	Applications
Plain carbon steel	Low alloy content, hardness up to 200 VPN. Ductile, tough deposit.	Build-up of ferritic carbon steels prior to surfacing.
Pearlitic steel	Higher carbon/manganese content, hardness 250 VPN. Ductile, tough deposit.	Build-up and repair of friction (metal to metal) wear (e.g. idler rolls on tracked vehicles).
Martensitic–pearlitic steel	Low-alloy ferritic steel with hardness up to 400 VPN. Poor ductility and toughness.	Friction and mildly abrasion resistant applications, e.g. crusher rolls.
Martensitic steel	High-alloy (C,Mn,Cr,Mo,B). Hardness up to 800. Not machinable, low ductility.	Low-impact abrasion situations. Cutting edges, and digger teeth.
Austenitic manganese–nickel	Carbon steel alloyed with manganese & nickel. Low hardness (200 VPN), good ductility as deposited, work hardens to 500 VPN.	Repair of austenitic manganese steel. Crusher hammers. Good impact resistance and fair surface abrasion properties.

continued

Appendix 5 (*Continued*).

Consumable type	Features	Applications
Austenitic stainless steel	Normally 308- or 309-type stainless steel. Very good ductility and toughness, good impact and corrosion resistance, low abrasion resistance.	Good buffer layer for high-impact loading, good wet-corrosion properties. Use as build up on gyratory crushers and heavy hammers. High cost.
Chromium carbide white irons	High chromium–carbon white iron. Hard carbide particles in austenitic, pearlitic or martensitic matrix. Very good abrasion resistance, poor impact, fair corrosion resistance. Hardness depends on matrix; from 350 to 800 VPN.	Abrasion-resistant surfacing of earth-moving parts. Abrasion resistance best with martensitic matrix, impact and corrosion best with austenitic matrix. Very flexible range.
Tungsten carbide white irons	Tungsten carbide in martensitic white-iron matrix. Very hard (up to 850 VPN) abrasion-resistant deposit, poor impact properties.	Fine abrasion-resistant applications; sand chutes and dredger buckets.
Nickel- and cobalt-based alloys	May contain carbides (e.g. tungsten carbide) in tough, ductile, corrosion-resistant matrix. Higher cost than iron-based.	Good high-temperature abrasion and erosion resistance. Valves and valve seats.

Note: VPN hardness scale; Vickers pyramid number.

Appendix 6
Plasma keyhole welding parameters

Material	Thickness (mm)	Current (A)	Orifice diameter (mm)	Plasma gas flow (l min^{-1})	Shield gas flow (l min^{-1})	Travel speed (m min^{-1})
Plain carbon	1.2	130	2.8	4.2	20	0.35
steel	3.2	270	3.2	2.8	10	0.23
	6.4	300	3.2	2.8	10	0.23
Austenitic	1.2	95	2.8	3.8	20	0.96
stainless	3.2	190	2.8	4.2	20	0.84
steel	6.4	240	3.5	8.5	24	0.35
	12.7	320	3.2	4.7	10	0.18
Nickel	3.2	200	2.8	4.7	20	0.69
	6.4	250	3.5	7.1	20	0.35
Copper	3.2	60	3.2	1.9	10	0.23
Incoloy	3.2	180	2.8	5.7	20	0.69
	6.4	240	3.5	7.1	20	0.35
Nimonic	1.6	135	2.8	3.5	20	0.92
	3.2	200	2.8	3.8	20	0.76

Appendix 7
Plasma keyhole welding of titanium

Thickness (mm)	Current (A)	Orifice diameter (mm)	Plasma gas flow (l min^{-1})	Shield gas flow (l min^{-1})	Trailing shield (l min^{-1})	Travel speed (m min^{-1})
1.6	132	2.8	3.8	20	50	0.92
3.2	185	2.8	3.8	20	50	0.50
4.8	190	2.8	5.6	20	60	0.35
6.4	245	2.4	5.6	20	70	0.23

References

[1] American Welding Society 1989 *Welding Handbook* (vol 1 *Welding Technology*) 8th edn
[2] British Standards Institution 1983 *BS 499: Part 1* (Welding Terms and Symbols, Glossary of welding, brazing and thermal cutting)
[3] Bay N 1986 Cold welding pts 1–3 *Metal Construction* **18** (6, 8, and 10)
[4] Benn B 1988 Friction welding of butt joints for high duty applications *Welding and Metal Fabrication* August/September
[5] Nicholas E D and Teale R A 1988 Friction welding of duplex stainless steel *Offshore Technology Conf. (Houston, Texas, 2–5 May 1988)*
[6] Nicholas E D 1982 A friction welding application in the nuclear power industry *Welding Inst. Res. Bull.* **23** (1)
[7] Essa A A and Bahrani A S 1989 The friction joining of ceramics to metals *Proc. Int. Conf. on the Joining of Materials, JOM-4 (Helsingor, Denmark, 19–22 March 1989)*
[8] Thomas W M *et al* 1984 Feasibility studies into surfacing by friction welding *TWI Res. Rep.* 236 (Cambridge: The Welding Institute)
[9] Nicholas D and Watts E 1990 Friction welding—a sparkling success *The Welding Institute, Connect* (8) April
[10] Bartle P M 1983 Diffusion bonding—principles and applications *Welding Inst. Res. Bull.* **24** (3)
[11] Johnson K I *et al* 1979 MIAB welding, principles of the process *Metal Construction* **11** (11)
[12] Edson D A 1983 Application of MIAB welding *Proc. Conf. Developments and Innovations for Improved Welding Production (The Welding Institute, Birmingham, England, 13–15 Sept 1983)*
[13] Smith D S 1989 Control of quality and cost in fabrication of high integrity pipework systems *Proc. 'Weldfab Midlands' Seminar (25–26 Oct 1989)*

[14] 1989 Data sheet 'Control of Welding—Welding Procedures', Basic Welding Data No 11 *Welding and Metal Fabrication* **57** (10)

[15] Salter G R 1970 Introduction to arc welding economics *Metal Construction and British Welding J.* June

[16] Various authors 1989 Improving productivity and the control of quality in welded fabrication, *'Weldfab Midlands' Seminar, Birmingham, England, 25–26 October 1989* (*Welding and Metal Fabrication*)

[17] 1989 Data sheet 'Safety in Welding and Cutting', Welding Data Sheet No 10 *Welding and Metal Fabrication* **57** (9)

[18] Puska 1989 Welding in the 1990s *Svetsaren* **2** (Esab Group)

[19] Anon 1990 What of the 90s *Welding Distributor News* January

[20] Sandham J 1987 A comparative study of synergic pulsed MIG and tubular electrode GMAW *MSc Thesis* School of Industrial Science, Cranfield Institute of Technology

[21] Weston J 1985 Automation and robotisation in welding—the UK scene *Proc. Conf. 'Automation and Robotisation in Welding and Allied Processes'* (*Strasbourg, 2–3 Sept 1985*) (Oxford: IIW/Pergamon) p 347

[22] Belforte D A 1990 Preview of 1990: the picture begins to clear *Industrial Laser Rev.* January

[23] Norrish J 1989 Arc welding power source designs to meet the needs of industry—Part 1 *Welding and Metal Fabrication* **57** (2)

[24] Manz A F 1969 Inductance vs slope control for gas metal arc power *Welding J.* September

[25] Essers W G 1979 New types of power sources for arc welding *International Institute of Welding Document* XII-F-204–79

[26] Kanervisto M and Pullinen J 1987 'Solid state techniques improve welding and safety *Welding and Metal Fabrication* June

[27] Grist J 1977 Solid state controls—what solid-state controls can do for welding *Welding Design and Fabrication* August

[28] Needham J C 1977 Transistor power supplies for high performance arc welding *Welding Inst. Res. Bull.* March

[29] Irving R R 1983 Power packed transistors arrive in metalworking *Iron Age* 15 June

[30] Rodrigues A 1981 Improving the efficiency of the series regulator welding power source *IEE Seminar* (*London, March 1981*)

[31] Lowery J 1978 A new concept for AC/DC power sources for TIG welding *4th Int. Conf. on Advances in Welding Processes* (*The Welding Institute, May 1978*)

[32] Kyselica S 1987 High frequency reversing arc switch for plasma arc welding of aluminium *Metal Construction* **19** (11)

[33] Colens A 1976 Electronic welder *RCA Power Semiconductor, Application Engineering Report* (*Liege, May 1976*)

[34] Norrish J 1991 Arc welding power sources—design evolution and welding characteristics *International Institute of Welding Document* XII-1215-91

[35] Judson P and McKeown D 1982 Advances in the control of weld metal toughness *Proc. 2nd. Int. Conf. on Offshore Welded Structures* (Cambridge: The Welding Institute) paper 3

[36] Horsfield A 1990 Advances in flux covered manual metal arc electrodes *Welding and Metal Fabrication* **58** (1)

[37] Anon 1989 Sahara ReadyPack *Smitweld Reportage* No 3 September

[38] Farrar J C M 1990 Developments in stainless steel welding consumables *Welding and Metal Fabrication* **58** (1)

[39] Thornton C E 1988 Submerged arc welding consumables for high toughness applications *Welding and Metal Fabrication* **56** (7)

[40] Fraser R *et al* 1982 High deposition rate submerged arc welding for critical applications *Proc. 2nd Int. Conf. on Offshore Welded Structures* (Cambridge: The Welding Institute) paper 12

[41] Troyer W and Mikurak J 1974 High deposition submerged arc welding with iron powder joint fill *Welding J.* **53** (8)

[42] Kohno A and Barlow J A 1982 Improved submerged arc productivity with metal powder additions *Welding Inst. Res. Bull.* December

[43] Paranhos R 1991 A numerical method for welding parameter prediction and optimisation *PhD Thesis* School of Industrial and Manufacturing Science, Cranfield Institute of Technology

[44] McKeown D 1990 Trends and developments in MIG welding consumables *Welding and Metal Fabrication* **58** (1)

[45] Lesnewich A 1955 Electrode activation for inert gas metal arc welding *Welding J.* **34** (12)

[46] Mistry H R 1986 Effects of wire and shielding gas composition on the toughness of pulsed MIG welds in HSLA steel *MSc Thesis* School of Industrial Science, Cranfield Institute of Technology

[47] Gustafsson B-O and Widgery D 1989 Cored wires—a review *Int. J. for the Joining of Materials* **1** (3)

[48] Rodgers K J and Lochhhead J C 1987 Self-shielded flux cored arc welding—the route to good fracture toughness *Welding J.* July

[49] Sandham J 1987 A comparative study of synergic pulsed MIG and tubular electrode GMAW *MSc Thesis* School of Industrial Science, Cranfield Institute of Technology

[50] Widgery D J 1988 Flux-cored wire: an update *Welding and Metal Fabrication* **56** (3)

[51] Yamamuchi N and Taka T 1979 TIG arc welding with hollow tungsten electrodes *International Institute of Welding Document* 212-452-79

[52] Allum C J 1987 Nitrogen absorption from welding arcs *International Institute of Welding Document* 212-659-86

[53] Norrish J and Alfaro S A unpublished

[54] Norrish J, Hilton D E and Mistry H R 1986 The effect of shielding gas composition in the MAG welding of line pipe *Proc. 3rd Int. Conf. on the Welding and Performance of Pipelines (The Welding Institute, London, 18–21 November 1986)* paper 19

[55] Stenbaka N and Svennson O 1987 Some observations of pore fomation in gas metal arc welding *Scand. J. Metall.* **16**

[56] Scheibner P 1979 Results of investigations on the formation of spatter during metal active gas arc welding in the GDR *International Institute of Welding Document* XII-B-271-79

[57] Norrish J 1974 High deposition MIG welding with electrode negative polarity *Proc. 3rd Conf. on Advances in Welding Processes* (Cambridge: The Welding Institute) paper 16

[58] Waering A J 1989 Cast to cast variation in austenitic stainless steel *MSc Thesis* Cranfield Institute of Technology

[59] Hilton D E 1982 Helium argon mixtures prove more economic than argon for aluminium welds *Welding and Metal Fabrication* June

[60] Kennedy C R 1970 Gas mixtures for welding *Aust. Welding J.* September

[61] Stenbacka N 1990 Shielding gas and cored wire welding characteristics—weld metal mechanical properties *AGA Innovation*

[62] Johnson K I *et al* 1979 MIAB welding—principles of the process *Metal Construction* **11** (11)

[63] Patchett B M 1978 MIG welding of aluminium with an argon–chlorine gas mixture *Metal Construction* **10** (10)

[64] Bicknell A C and Patchett B M 1985 GMA welding of aluminium with argon/freon shielding gas mixtures *Welding J.* May

[65] Heiple C R and Burgardt P 1985 Effects of SO_2 shielding gas additions on GTA weld shape *Welding J.* June 159s–162s

[66] Stenbacka N 1989 Reduction of ozone using Mison shielding gases—a few basic facts *AGA AB Innovation*

[67] Anon 1990 The problem of ozone in TIG welding *AGA Gas Division Report* GM116e (AGA: Sweden)

[68] Anon 1990 Facts about: ozone reduction with Mison shielding gases *AGA Report* (AGA: Sweden)

[69] Beck Hansen E 1988 *Emissions of Fume, Ozone and Nitrous Gases During MAG Welding of Low Alloy Steel and Stainless Steel, 1987* (Appendix with results of MAG welding with shielding gas $75Ar/25CO_2$ and Mison 25) (The Danish Welding Institute)

[70] Imaizumi H and Church J 1990 Welding characteristics of a new

welding process, T. I. M. E. process *International Institute of Welding Document* XII-1199-90

[71] Kirkpatrick I 1989 Welding with lasers *Welding and Metal Fabrication* August/September

[72] Norris I M 1988 Optimisation of gas jet plasma control system for laser welding with a 10 kW CO_2 laser *Welding Institute Research Report* TWI 368/1988

[73] Brown M J 1976 Initiation of gas tungsten arcs by high voltage DC *Welding Institute Research Report* 31/1976/P

[74] Biegelmeier G 1985 *Effects of Current Passing Through the Human Body and the Electrical Impedance of the Human Body* (Berlin: VDE)

[75] Edberg H 1988 Improved touch starting techniques for TIG welding *Metal Construction* **20** (3)

[76] Nixon J H TIG touch striking trials *Private Communication*

[77] Melton G B and Street J A 1982 Piezoelectric techniques for the initiation of TIG arcs *Welding Inst. Res. Bull.* **23** (2)

[78] Melton G B 1980 Initial current characteristics of welding sets for gas tungsten arc welding *Welding Institute Research Report* 7729.01/80/240.2

[79] Goldman K 1965 An evaluation of electrodes for DC TIG welding *BOC Ltd Internal Report*

[80] Masumoto I, Matsuda F and Ushio M 1988 Development and application of new oxide activated tungsten electrodes in Japan *International Institute of Welding Document* XII-1047-87

[81] Rudaz A 1979 Summary report on the use of pulsed currents in the TIG process *International Institute of Welding Document* XII-F-201-79 (also published in *Welding in the World* **18** (3/4) 1980)

[82] Boughton P and Males B O 1970 Penetration characteristics of pulsed TIG welding *Welding Institute Research Report* R/RB/P43/70

[83] Shimada W and Gotoh T 1976 Characteristics of high frequency pulsed DC TIG welding process *International Institute of Welding Document* XII-628-76

[84] Grist F J 1975 Improved lower cost aluminium welding with solid state power source *Welding J.* **54** (5)

[85] Lowery J A 1978 New concept for AC/DC power sources for TIG welding *Welding Institute Conf. 'Advances in Welding Processes' (Harrogate, May 1978)* paper 7

[86] Maruo H and Hirata Y 1986 Rectangular wave AC TIG welding of aluminium alloy *International Institute of Welding Document* 212-647-86

[87] Kyselica S 1987 High frequency reversing arc switch for plasma arc welding of aluminium *Metal Construction* **19** (11)

[88] Tomsic M J and Barhorst S W 1983 Applications of keyhole plasma arc welding of aluminium, GTAW cathode etching of aluminium tubes and dabber TIG process *Proc. 31st Ann. Conf. of The Australian Welding Institute (Sydney, 16–21 October 1983)*

[89] Manz A F 1977 Hot wire welding and surfacing techniques *Welding Research Council Bulletin* AWS

[90] Saenger J F and Manz A F 1968 High deposition gas tungsten arc welding *Welding J.* May

[91] Wright J P 1989 High frequency phase controlled power regulator for hot wire TIG welding *MSc Thesis* School of Industrial Science, Cranfield Institute of Technology

[92] Schultz J P 1980 New possibilities in plasma-arc and dual shielding gas TIG welding *Conf. on Welding Fabrication and Surface Treatment (Singapore 15–18 July 1980)* (Singapore Welding Society)

[93] Richardson I M 1989 Plasma welding—the current status *Welding and Metal Fabrication* **57** (5)

[94] Woodford D and Norrish J 1975 Progress in the application of the plasma process. *The Welding Institute Conf. 'Exploiting Welding in Production Technology' (London, 22–24 April 1975)*

[95] Wealleans J W and Allen B 1969 Towards automating the TIG process *Welding and Metal Fabrication* March

[96] Anderson J E and Yenni D M 1970 Multi-cathode gas tungsten arc welding (Union Carbide ref. 52-538) *Welding J.*

[97] Savage W F, Nippes E F and Agusa K 1979 Effect of arc force on defect formation in GTA welding *Welding J.* July

[98] Goldman K 1963 Electric arcs in argon:heat distribution *Br. Welding J.* July

[99] Spiller K R and MacGregor G J 1970 Effect of electrode vertex angle on fused weld geometry in TIG welding *Proc. Conf. on Advances in Welding Processes (The Welding Institute 14–16 April 1970)*

[100] Key J F 1980 Anode/cathode geometry and shielding gas inter-relationships in GTAW *Welding J.* December

[101] Mills K C and Keene B J 1989 The factors affecting variable weld penetration *International Institute of Welding Study Group 212* August

[102] Keene B J 1988 The effect of thermocapillary flow on weld pool profile *NPL Report* DMA(A) 167 (Teddington, UK: National Physical Laboratory)

[103] Waering A J 1988 Control of TIG welding in practice *Seminar 'Who's in Control'* (*Cranfield Institute of Technology May 1988*)

[104] Smal C 1988 Oscillations in voltage and spectral radiation signals from GTAW arcs *MSc Thesis* School of Industrial Science, Cranfield Institute of Technology

[105] Pratt N 1989 Fluctuations in voltage and light signals from GTAW arcs *MSc Thesis* School of Industrial Science, Cranfield Institute of Technology

[106] Needham J C, Cooksey C I and Milner D R 1960 The transfer of metal in inert gas shielded arc welding *Welding J.* **7**

[107] Ma J 1982 Metal transfer in MIG welding *PhD Thesis* Cranfield Institute of Technology

[108] Norrish J and Ward B Metal transfer analysis of FCAW using an Imacon camera (unpublished)

[109] Norrish J and Richardson I M 1988 Metal transfer mechanisms *Welding and Metal Fabrication* **56** (1)

[110] van Adrichem Th J 1969 Metal transfer *International Institute of Welding Document* 212-171-69

[111] Anon 1976 Classification of metal transfer XII-636-76

[112] Norrish J 1974 High deposition MIG—welding with electrode negative polarity *Proc. 3rd. Conf. on Advances in Welding Processes* (Cambridge: The Welding Institute) paper 16

[113] Stevenson A W 1988 Analysis of burn-off behaviour of flux cored arc welding *MSc Thesis* Cranfield Institute of Technology

[114] Bucknall P W 1990 Influence of power source dynamics on the burn-off rate during pulsed GMA welding *MSc Thesis* Cranfield Institute of Technology

[115] Chang C H 1989 Investigation of arc characteristics and metal transfer for flux cored electrode in GMA welding *MSc Thesis* Cranfield Institute of Technology

[116] Blackman S and Norrish J 1988 Pulsed MIG welding with gas shielded flux cored wires. Synergic MIG update *Welding and Metal Fabrication* April

[117] Boughton P and MacGregor G 1974 Control of short circuiting in MIG-welding *Welding Res. Int.* **1.4** (2)

[118] Manz A F 1968 One knob welder *Welding J.* **47** (9)

[119] Norrish J 1972 Developments in MIG welding of sheet steel and strip *Sheet Metal Industries (Proc. Conf. on Sheet Steel and Strip Welding (Kenilworth, England, 15–17 March 1972))* (Birmingham: The Welding Institute) p 6

[120] Amin M and Watkins P V C 1977 Synergic pulse MIG welding *Welding Institute Research Report* 46/1977/P

[121] International Institute of Welding 1985 Synergic power supplies—

classification and questionnaire *International Institute of Welding Document* XII-905-85

[122] Lowery J 1975 Electronic control systems in welding *Electro Technology* **98** (10)

[123] Amin M and Naseer A 1987 Synergic control in MIG welding, 2—Power—current controllers for steady DC open arc operation *Metal Construction* **19** (6)

[124] Quigley M B C 1986 High power density processes *The Physics of Welding* (Oxford: IIW/Pergamon) p 306

[125] Metcalfe J C and Quigley M B C 1975 Heat transfer in plasma arc welding *Welding J.* March

[126] Woodford D R 1972 Plasma welding data sheets *BOC Guidance Notes*

[127] Bland J 1973 *Recommended Practices for Plasma-arc Welding* AWS C5.1-73 (American Welding Society)

[128] Pinfold B E 1974 Plasma arc welding—part 4 Welding technology data sheet *Welding and Metal Fabrication*

[129] Lavigne D, Van Der Have P and Maksymowicz M 1988 Automatic plasma arc welding *Joining and Materials* July

[130] Zhou Yin 1985 The influence of focusing gas flow on working stability of high capacity plasma torch *Trans China Weld Inst.* **6** (2)

[131] Pattee H E, Meister R P and Monroe R E 1968 Cathodic cleaning and plasma arc welding of aluminium *Welding J.* May

[132] Smart M D and Pinfold B E 1971 Comparison of modulated and unmodulated current plasma welding *Welding and Metal Fabrication* September

[133] Lucas W 1978 Pulsed plasma welding *Welding Inst. Res. Bull.* **19**

[134] Narita K *et al* 19XX Plasma arc welding of pipelines: a study to optimise welding conditions for horizontal fixed joints of mild steel pipes *Int. J. Pressure Vessels and Piping*

[135] Miller H R and Filipski S P 1966 Automated plasma arc welding for aerospace and cryogenic fabrications *Welding J.* June

[136] Woolcock A and Ruck R J 1978 Keyhole plasma arc welding of titanium plate *Metal Construction* December

[137] Wu Chih Chiang and Pinfold B E 1979 Operational envelopes for plasma keyhole welding of titanium *Welding and Metal Fabrication* November

[138] Tomsic M J and Barhorst S W 1983 Applications of keyhole plasma arc welding of aluminium, GTAW cathode etching of aluminium tubes and dabber TIG process *Proc. 31st. Ann. Conf. of The Australian Welding Institute* (*Sydney, 16–21 October 1983*)

[139] Herziger G 1985 The influence of laser induced plasma on laser materials processing *Annual Review of Laser Processing, Industrial Laser Handbook*

[140] Matsunawa A, Yoshida H and Katayama S 1985 Beam plume interaction in pulsed YAG laser processing *International Institute of Welding Document* 212-617-85

[141] Megaw J H P C, Hill M and Johnson R 1981 Laser welding of steel plates with unmachined edges *UKAEA Report* CLM-P649

[142] Banas C M 1979 *US Patent* No 4152575

[143] Arata Y *et al* 1985 Fundamental phenomena in high power CO_2 laser welding *Trans. JWRI* **14** (1)

[144] Shinmi A *et al* 1985 Laser welding and its applications for steel making process *Laser Welding, Machining and Materials Processing (Proc. Int. Conf. on Applications of Lasers and Electro-Optics, ICALEO'85)* (IFS Publications)

[145] Anon 1989 Laser welded parts give greater speed and accuracy *Engineering Lasers* October

[146] Roessler M D 1990 A fresh look at lasers in automotive applications *Industrial Laser Rev.* February

[147] Metzbower E A 1983 Laser welding of mild steel at NIROP Minneapolis (USA) *Metal Construction* **15** (10)

[148] Mazumder J and Steen W M 1980 Welding of Ti-6A1-4V by a continuous wave CO_2 laser *Metal Construction* **12** (9)

[149] Matsuda J *et al* 1988 TIG or MIG arc augmented laser welding of thick mild steel plate *Joining and Materials* **1** (1)

[150] Steen W M and Eboo M 1979 Arc augmented laser welding *Metal Construction* **11** (7)

[151] Weedon T M, Burrows G and Thompson P G 1989 Nd-YAG lasers in car body cutting and welding *ISATA Symp. (Florence, 1989)* (Rugby, UK: Lumonics)

[152] Ireland C L M 1990 Current state of power laser technology and future trends *Manufacturing Technology—Europe 1990*

[153] Hetcht J 1990 *Understanding Lasers* (Indiana: Howard W Sams)

[154] Sato *et al* 1989 Experience with CO lasers *Proc. 6th. Int. Conf. 'Lasers in Manufacturing' (May 1989)*

[155] Quigley M B C 1986 High power density welding *The Physics of Welding* (Oxford: IIW/Pergamon) p 306

[156] Benn B 1988 Joining technology at Rolls Royce PLC, Bristol *Welding and Metal Fabrication* **56** (6)

[157] Nightingale K R 1983 Electron beam welding of copper and its alloys *Welding Inst. Res. Bull.* **24** (1)

[158] Russell J D 1981 Electron beam welding—a review *Metal Construction* **13** (7)

[159] Malin V Y 1983 The state-of-the-art of narrow gap welding *Welding J.* **62** (6)

[160] Farish E 1989 Modelling the GTAW process with cold wire addition *MSc Thesis* Cranfield Institute of Technology

[161] Randall M D and Nelson J W 1979 CRC automatic welding system *Proc. Conf. Recent Developments in Pipeline Welding* (Cambridge: The Welding Institute)
[162] Meister R P *et al* 1966 Narrow gap welding process *Br. Welding J.* **13** (5)
[163] Modenesi P 1990 Statistical modelling of narrow gap gas metal arc welding process *PhD Thesis* Cranfield Institute of Technology
[164] Kimura S *et al* 1979 Narrow gap gas metal arc welding process in the flat position *Welding J.*
[165] Alfaro S 1989 Mathematical modelling in narrow gap submerged arc welding *PhD Thesis* Cranfield Institute of Technology
[166] Alfaro S C A and Apps R L A 1989 Mathematical modelling of narrow gap submerged arc welding *Proc. Int. Conference on the Joining of Materials, JOM-4 (Helsingør, Denmark, 19–22 March 1989)*
[167] de Altamer A 1980 Submerged arc narrow gap welding *Metal Construction* **12** (10)
[168] Hirai *et al* 1981 Application of narrow gap submerged arc welding to fabrication of 2.25%Cr–1%Mo forged steel heat exchangers *International Institute of Welding Document* XIIA-009-81
[169] Lochead J C 1983 Narrow gap welding *Proc. Joint Conf. South African Institute of Welding and Institute of Mechanical Engineers 'Welding and the Engineer, the Challenge of the 80's'*
[170] Kennedy N A 1986 Narrow gap submerged arc welding of steel; Part 1 Applications *Metal Construction* **18** (11)
[171] Ellis D J 1988 Mechanised narrow gap welding of ferritic steel *Joining and Materials* **1** (2)
[172] Curtis G, Lovegrove G L and Farrar R A 1990 'WASPS'—an advisory system for welding process selection *Welding and Metal Fabrication* **58** (1)
[173] Hicks J 1979 *A Guide to Designing Welds* (Cambridge: Abington Publishing)
[174] British Standards Institution 1986 *BS 4871* (Specification for approval testing of welders working to approved welding procedures); 1986 *BS 4872* (Specification for approval testing of welders when welding procedure approval is not required)
[175] Queen D M 1989 The evaluation and implementation of a computerised welding procedure database management system *Shell Expro*
[176] Robinson P *Private communication*
[177] British Standards Institution 1979 *BS 638: Part 2*—(Arc welding power sources, equipment and accessories—part 2. Specification for air cooled power sources for manual metal-arc welding with covered electrodes and for TIG welding)

[178] Smith D S 1988 Control of MMA welding in practice—a user's view *Who's In Control Seminar* (*Cranfield Institute of Technology, 25–26 May 1988*)

[179] Smith C J 1988 The calibration of welding equipment *Who's in Control Seminar* (*Cranfield Institute of Technology, 25–26 May 1988*)

[180] Norrish J 1988 Computer based instrumentation for welding *Proc. Conf. Computer Technology in Welding* (*The Welding Institute, Cambridge, 8–9 June 1988*)

[181] Nixon J H N 19XX *Instrumentation for Arc Welding* (Cranfield: Cranfield Institute of Technology)

[182] Chawla K C 1991 *A General Purpose PC Based Welding Data Logger* (Cranfield: Cranfield Institute of Technology)

[183] Kuszleyko R 1979 An objective solution of the welding process stability evaluation *International Institute of Welding Document* XII-F208-79

[184] Duncan A 1980 Evaluation of power sources for manual metal arc welding by use of stability criteria *MSc Thesis* University of Aston, Birmingham

[185] Carrer A 1960 Dynamic behaviour of DC generators for arc welding *Br. Welding J.* January

[186] Buki A and Gorenshtein I M 1967 The stability and self adjustment of metal deposition with short circuits of the arc gap *Automatic Welding* 11

[187] Lucas W and Butler M 1981 An evaluation of minicomputer techniques for data acquisition and analysis in arc welding process research *Welding Institute Research Report* 134/1981

[188] Batson M S 1981 Effect of gas composition on arc stability during the short circuit welding of stainless steel *MSc Thesis* University of Aston, Birmingham

[189] Norrish J, Holmberg L and Hilton D E 1989 Optimisation of argon, oxygen carbon dioxide gas mixtures for the GMA welding of plain carbon steel *Proc. Int. Conf. on the Joining of Materials, JOM-4* (*Helsingør, Denmark 19–22 March 1989*)

[190] Weinschenk H E and Schellhase M 1979 Statistical analysis of arcing voltage in the CO_2 shielded welding process *Proc. Conf. Arc Physics and Weld Pool Behaviour* (*The Welding Institute, London, 8–10 May 1979*)

[191] Gupta S R, Gupta P C and Rehfeldt D 1988 Process stability and spatter generation during dip transfer in MAG welding *International Institute of Welding Document* 212-710-88

[192] Shinoda T, Kaneda H and Takeuchi Y 1989 An evaluation of short circuiting phenomena in GMA welding *Welding and Metal Fabrication* 57 (10)

[193] Mita T, Sakabe A and Yokoo T 1988 Quantitative estimates of arc stability for CO_2 gas shielded arc welding *Welding Int.* (2)

[194] Budai P 1988 Measurement of drop transfer stability in weld processes with short circuiting drop transfer *International Institute of Welding Document* 212-711-88

[195] Andrews D R and Broomhead J H W 1975 Quality assurance for resistance spot welding *Welding J.* June

[196] Broomhead J H W and Dony P H 1990 Resistance spot welding quality assurance *Welding and Metal Fabrication* **58** (6)

[197] Rashid O A F, Hodgson D C and Broomhead J H W 1983 A microprocessor based monitor for the resistance welding of mild steel and aluminium *Proc. Int. Conf. 'Development and innovation for improved welding production' (Birmingham, UK, September 1983)*

[198] Janota M and Kuban J 1985 Adaptive systems of process control for spot welding robot cells *Proc. Conf. Automation and Robotisation in Welding and Allied Processes (Strasbourg, 1985)* (Oxford: Pergamon)

[199] Crookall J R and Philpott M L 1984 Direct arc sensing adaptive quality control of MIG welding robots *Proc. 3rd. SERC Robotics Initiative Conference (University of Surrey)*

[200] Philpott M L 1986 Direct arc sensing for robot arc welding *PhD Thesis* Cranfield Institute of Technology

[201] Needham J C and Bourton M 1988 Monitoring and control techniques for arc welding *Who's in Control Seminar (Cranfield Institute of Technology, 25–26 May 1988)*

[202] Brown K W 1988 Programmable error monitors—a new breed of QA watchdogs *Welding Inst. Res. Bull.* **29** (3/4)

[203] Philpott M L and Norrish J 1988 Process monitoring and statistical process control *Who's in Control Seminar (Cranfield Institute of Technology, 25–26 May 1988)*

[204] Papritan J C and Helzer S C 1991 Statistical process control for welding *Welding J.* **70** (3)

[205] Thomas M A 1987 The implementation and benefits of statistical process control *Quality Assurance* **13** (1)

[206] Shewhart W A 1931 *Economic Control of Quality of Manufactured Products* (New York: Van Nostrand)

[207] Blackmon D R and Kearney F W 1983 A real time quality approach to quality control in welding *Welding J.* August

[208] Anon 1989 Destructive testing of wheel weldments cut in half *Welding J.* April

[209] Reilly R 1991 Real-time weld quality monitor controls GMA welding *Welding J.* **70** (3)

[210] Evans C R C 1988 Monitoring the welding process *The Fabricator* November (Rockford, IL: FMA)

[211] British Standards Institution 1984 *BS 5750 Part 1* (Quality systems—Specification for design, manufacture and installation. Clause 4.9.2)

[212] Ogata K 1984 *Modern Control Engineering* (Prentice Hall)

[213] Åström K J and Wittenmark B 1980 *Adaptive Control* (Addison Wesley)

[214] Richardson R W 1986 Robotic weld joint tracking systems—theory and implementation methods *Welding J.* November

[215] Philpott M L and Crookall J R 1986 Seam tracking by direct arc sensing *Manufacturing Systems Engineering* (DDMS, Cranfield Institute of Technology)

[216] Skjølstrup C E 1990 The introduction of robots in a shipyard *Welding Rev.* **9** (1)

[217] Kohn M L *et al* 1985 A robotic cell for welding large aluminium structures using a two-pass optical system *Proc. First Int. Conf. on Advanced Welding Systems (The Welding Institute, London, 19–21 November 1985)*

[218] Hughes R 1985 Arc guided robot plasma welding *Proc. First Int. Conf. on Advanced Welding Systems (The Welding Institute, London, 19–21 November 1985)* paper 17

[219] Nomura H *et al* 1984 The development of automatic TIG welding process with seam tracking *Proc. Int. Conf. on Quality and Reliability in Welding (Hangchou, China, 6–8 September 1984)*

[220] Tan C and Lucas J 1986 Low cost sensors for seam tracking in arc welding *Proc. 1st. Int. Conf. 'Computer Technology in Welding' (The Welding Institute, London, 3–5 June 1986)*

[221] Drews P and Starke G Development approaches for advanced adaptive control in automated arc welding *Proc. Conf. Automation and Robotisation in Welding and Allied Processes (Strasbourg 1985)* (Oxford: Pergamon)

[222] Barratt J W and Davey P G 1987 New horizons for laser stripe sensors *Metal Construction* **19** (12)

[223] Agapakis J E *et al* 1986 Joint tracking and adaptive robotic welding using vision sensing of the weld joint geometry *Welding J.* November

[224] Richardson R W *et al* 1984 Coaxial arc weld pool viewing for process monitoring and control *Welding J.* **63** (3)

[225] Richardson R W and Corardy C C 1990 Coaxial vision based GMAW: initial design and test. Research report summary *Edison Welding Inst. Insights* **4** (1)

[226] Stroud R R and Harris T J 1990 Seam tracking butt and fillet welds using ultrasound *Joining of Materials* **2** (1)

[227] Anon 1986 *Composition Controlled Sensing Technology* (Rockville, MD: U-Weld Automation)

[228] Habil F, Erdmann-Jesnitzer E, Feustel E and Rehfeldt D 1966 Acoustic investigations of the welding arc *International Institute of Welding Document* 212-86-66

[229] Rommanenkov E I *et al* 1976 Control of arc length on the basis of its spectral radiation *Welding Production* 9

[230] Boughton P, Rider G and Smith C J 1978 Feedback control of weld penetration *Proc. Conf. Advances in Welding Processes (Cambridge, UK 1978)*

[231] Peters C N D 1987 The use of backface penetration control methods in synergic pulsed MIG welding *MSc Thesis* Cranfield Institute of Technology

[232] Watson D G 1986 The use of backface penetration control methods in synergic pulsed MIG welding *MSc Thesis* Cranfield Institute of Technology

[233] Naseer A and Lucas W 1988 Evaluation of a video system for control of weld bead penetration in TIG welding *Welding Institute Research Report* 357/1988

[234] Salter R J and Deam R T 1987 A practical front face penetration control system for TIG welding *Proc. Conf. Developments in Automated and Robotic Welding (Welding Institute, London, 17–19 November 1987)*

[235] Madigan R B *et al* 1986 Computer based control of full penetration GTA welding using pool oscillation sensing *Proc. Conf. Computer Technology in Welding (Welding Institute, London, 1986)*

[236] Lucas W and Mallett R S 1975 Automatic control of penetration in pulsed TIG welding *Welding Institute Research Report* P/72/75

[237] Rokhlin S I 1989 In process radiographic control of arc welding *Materials Evaluation* **47** (3)

[238] Boillot J P *et al* 1985 Adaptive welding by fibre optic thermographic sensing: an analysis of thermal and instrumental considerations *Welding J.* July

[239] Bangs R E 1986 AI in the development of adaptive controls for fusion welding *SME Conf. Ultratech—Artificial Intelligence (Long Beach, CA, 22–25 September 1986)*

[240] Wareing A J, Murdock J and Sullivan P D 1988 Fabricating stainless steel cell liner plates. Synergic MIG update supplement *Welding and Metal Fabrication* April

[241] Anon 1987 Simple mechanisation enhances shipbuilding productivity *Metal Construction* **19** (3)

[242] Anon 1990 Automated welding with columns, booms and tractors *Welding and Metal Fabrication* **58** (2)

[243] Reeves K D and Miller C J 1985 The development of a remotely

controlled welding system for the site welding of AGR fuel standpipe closures *Int. Conf. on Advanced Welding Systems (Welding Institute, London, November 1985)*

[244] Carrick L, Hick A B, Salmon S and Wareing A J 1985 A new welding technique for stainless steel pipe butt welds *Metal Construction* **17** (6)

[245] Halford P 1987 *Productivity and Quality Benefits From Improved Pipework Installation Techniques with Reference to The Vitrification Plant* (British Nuclear Fuels)

[246] *UK Patent* 8411823

[247] Adams D 1979 *The Hitchhikers Guide To The Galaxy* (Pan)

[248] Archer J R and Blenkinsop P T 1986 Actuation for industrial robots *Proc. Inst. Mech. Eng.* **200** (B2)

[249] Kunzig L A *et al* 1985 Flexible resistance welding line of side panels in railway wagon assembly *Int. Conf. on Advanced Welding Systems (Welding Institute, London, November 1985)*

[250] 1988 Tipo the product of the system *Fiat Publication* 10067-1/88 (Turin: Fiat)

[251] Wright R R 1990 The modular approach to automated and mechanised welding *Welding and Metal Fabrication* March

[252] Knagenhjelm H O, Morris A W, Pinches C A, Bellis G A and Gjermundsen K 1985 The development of a mechanised welding system for deep waters *Int. Conference on Advance Welding Systems (Welding Institute, London, November 1985)*

[253] Berge J O, Håbrekke T and Knagenhjelm H O 1991 Automation in underwater hyperbaric pipeline welding *OMAE Conf. (Stavanger, Norway, 23–28 June 1991)* paper ASME-91-844

[254] Smith C J, Morgan-Warren E J and Salter R 1987 The development of a remote tube welding system for sealing containers *Proc. Conf. Developments in Automated and Robotic Welding (Welding Institute, London, 17–19 November 1987)*

[255] Smith C J, Morgan-Warren E J and Salter R 1987 The development of a microprocessor controlled MIG welding system for joining toxic components *Proc. Conf. Developments in Automated and Robotic Welding (Welding Institute, London, 17–19 November 1987)*

[256] Smith R and Williams A 1985 Flexible system expands possibilities for robotic welding *Metal Construction* **17** (5)

[257] Pekkari B 1989 Towards FMS: developments in arc welding processes and equipment for robotic applications *Exploiting Robots in Arc Welded Fabrication* (The Welding Institute) p 119

[258] Smith P A 19XX An example of FMS robot arc welding *Exploiting Robots in Arc Welded Fabrication* (The Welding Institute)

[259] Skjølstrup C E 1990 The introduction of robots in a shipyard *Welding Rev.* **9** (1)

[260] Norrish J and Gourd L M 1987 Evaluation of opportunities for automation in arc welding—a training package *Proc. Conf. on Developments in Automated and Robotic Welding* (*Welding Institute, London, 17–19 November 1987*)

[261] Fekken U 1987 Straightforward solutions to straightforward automatic welding jobs *Proc. Conf. on Developments in Automated and Robotic Welding* (*Welding Institute, London, 17–19 November 1987*)

[262] Tanner W R 1978 *Selling the Robot: Justification for Robot Installations* (Michigan: Society of Manufacturing Engineers)

[263] Malin V 1986 Problems in the design of integrated welding automation—Part I: Analysis of welding-related operations as objects for welding automation *Welding J.* November

[264] Anon 1987 UK robot take up falls with support *Engineering News* (24) February

[265] Challis H 1987 Robots out of Wonderland *Engineering News* (32) October

[266] Huang P Y and Sakurai M 1989 An assessment of factory automation in Japan: a general mail survey *Manufacturing Rev.* **2** (3)

[267] Davis K and Martin R 1990 *Industrial Society Meeting* (*February 1990*) ('Automation is setting worker against worker', the *Guardian* 23 February 1990)

Index